QUICK CALCULUS

QUICK CALCULUS

For Self-Study or Classroom Use

DANIEL KLEPPNER, Ph.D.
Professor of Physics
Massachusetts Institute of Technology

NORMAN RAMSEY, Ph.D.
Professor of Physics
Harvard University

JOHN WILEY & SONS

New York • Chichester • Brisbane • Toronto

Library of Congress Catalog Card Number: 65-20764
ISBN: 0-471-49112-8

Printed in the United States of America.

30 29 28 27 26 25 24 23 22 21

PREFACE

Before you plunge into *Quick Calculus* perhaps we ought to tell you what it is supposed to do. *Quick Calculus* should teach you the elementary techniques of differential and integral calculus with a minimum of wasted effort on your part; it is designed for you to study by yourself. Since the best way for anyone to learn calculus is to work problems, we have included many problems in this book. You will always see the solution to your problem as soon as you have finished it, and what you do next will depend on your answer. A correct answer generally sends you to new material, while an incorrect answer sends you to further explanations and perhaps another problem.

We hope that this book will be useful to many different people. The idea for it grew out of the problem of teaching college freshmen enough calculus so that they could start physics without waiting for a calculus course in college. However, it soon became apparent that the book would be useful in many other ways. For instance, both graduate and undergraduate students in economics, business, medicine and the social sciences need to use some elementary calculus. Many of these students have never taken calculus, or want to review the course they did take; they should be able to put this book to good use. Ambitious high school students who want to get a head start on their college studies should find *Quick Calculus* just the thing. Unlike most calculus texts, it emphasizes technique and application rather than rigorous theories, and is therefore particularly suited for introducing the subject. Beginning calculus students who want a different and simpler view of the subject should find the book helpful either for self-instruction or for classroom use. We particularly hope that this book will be of use to those people who simply want to learn calculus for the fun of it.

Because of the variety of backgrounds of those who will use this book, we start with a review of some parts of algebra and trigonometry which are useful in elementary calculus. If you remember your high school preparation in these subjects you will sail through this material in little time, whereas if you have had little math, or have long been away from math, you will want to spend more time on this review. As you will see, one of the virtues of the book is its flexibility — the time you spend on each portion depends on your particular needs. We hope that this will save you time so that you will find the book's title appropriate.

Daniel Kleppner
Norman Ramsey

Harvard University
Cambridge, Massachusetts

CONTENTS

Appendix A.   DERIVATIONS

Appendix B.   ADDITIONAL TOPICS

REVIEW PROBLEMS

TABLES

INDEX

QUICK CALCULUS

CHAPTER I

A FEW PRELIMINARIES

In this chapter the plan of the book is explained, and some elementary mathematical concepts are reviewed. By the end of the chapter you will be familiar with

(1) the definition of a mathematical function;

(2) graphs of functions;

(3) the properties of the most widely used functions: linear and quadratic functions, trigonometric functions, exponentials and logarithms.

1

In spite of its formidable name, calculus is not a particularly difficult subject. Of course you won't become a master in it over-night, but with diligence you can learn its basic ideas fairly quickly.

This manual will get you started in calculus. After working through it you ought to be able to handle many problems and you should be prepared to learn more elaborate techniques if you need them. But remember that the important word is *working*, though we hope you find that much of the work is fun.

Most of your work will be answering questions and doing prob-lems. The particular route you follow will depend on your answers. Your reward for doing a problem correctly is to go straight on to new material. On the other hand, if you can't do the problem cor-rectly, the solution will usually be explained and you will get additional problems to see whether you have caught on. In any case, you will always be able to check your answers immediately after doing a problem.

Many of the problems have multiple choice answers. The pos-sible choices are grouped like this: $\boxed{a \mid b \mid c \mid d}$. Choose an answer by circling your choice. The correct answer is always located at the bottom of the following page. Some questions must be answered with written words. Space for these is indicated by a blank, and you will be referred to another frame for the correct answer.

If you get the right answer but feel you need more practice simply follow the directions for the wrong answer. There is no premium for doing this book in record time.

Go on to frame 2.

2

In case you want to know what's ahead, here is a brief out-line of the book: this first chapter is a review which will be useful later on; Chapter II is on Differential Calculus and Chapter III is on Integral Calculus. The last Chapter, IV, contains a concise outline of all the earlier work. There are two Appendices — one giving formal proofs of a number of relations we use in the book and the other discussing some supplementary topics. In addition, there is a list of extra problems, with answers, and a section of Tables you may find useful.

A word of caution about the next few frames. Since we must start with some definitions, the first section has to be a good deal more formal than most other parts of the book.

First we will review the definition of a function. If you are already familiar with this, and with the idea of independent and dependent variables, you should skip on to Section 2 (frame 14). (In fact, in this chapter there is ample opportunity for skipping if you already know the material. On the other hand, some of the material may be new to you, and a little time spent on review can be a good thing.)

Go to 3.

3

    The definition of a function makes use of the idea of a *set*. Do you know what a set is? If so, go to 4. If not, read on.

    A *set* is a collection of objects — not necessarily material objects — described in such a way that we have no doubt as to whether a particular object does or does not belong to the set. A set may be described by listing its elements. Example: the set of numbers, 23, 7, 5, 10. Another example: Mars, Rome and France.

    We can also describe a set by a rule, for example: all the even positive integers (this set contains an infinite number of objects). Another set defined by a rule is the set of all planets in our solar system.

    A particularly useful set is the set of all real numbers, which includes all positive or negative real numbers such as 5, −4, 0, 1/2, −3.482, $\sqrt{2}$, etc., but does not include numbers which involve $\sqrt{-1}$.

    The mathematical use of the word "set" is similar to the use of the same word in ordinary conversation, as "a set of golf clubs".

Go to 4.

4

    In the blank below, list the elements of the set which consists of all the odd positive integers less than 10.

_____

Go to 5 for the correct answer.

5

    Here are the elements of the set of all odd positive integers less than 10:

    1, 3, 5, 7, 9.

Go to 6.

6

Now we are ready to talk about functions. Since the precise definition has to be rather formal, let's start with a simple illustration.

In some newspapers we can find a list of the temperature at each hour of the day. With such a list one particular temperature is associated with each hour of the day. In mathematics such an association between two different objects or quantities is called a *function*.

We will use the following formal definition of a function. (If you have already learned quite a different definition, see Appendix B1 pages 262–263.)

> If each element of a set A is associated
> with exactly one element of set B, then
> this association is called a *function* from
> A to B. The set A is called the *domain*
> of the function.

Go to 7.

7

Often we use a symbol, such as x, to represent any element of the set A (the domain of the function). The symbol x is then called the *independent variable*. If the symbol y represents the element of the set B associated by the function with the element x, we call y the *dependent variable*. These definitions are really quite reasonable: in a typical use of a function we first arbitrarily and independently select a specific value for the independent variable x and use the function to give us the value of the dependent variable y which is determined by (or *depends* on) x.

Go to 8.

8

Let's see if these definitions are clear by reconsidering the example in frame 6 of the temperature listed for each hour of the day.   Write in the correct words in the blank spaces of the following sentence.

The association between the temperature and the time in this listing is a _____ from the hour to the temperature.   If the symbol h is used to represent the hour of the day and the symbol T is used to represent the temperature, the independent variable is _____ and the dependent variable is _____ .

Go to 9 for the correct answers.

9

You should have written that the association is a *function* from the hour to the temperature.   The independent variable is h and the dependent variable is T.

If you got all of these correct, you have caught on to the definitions and should skip to frame 11.   If you missed any, you should go to frame 10.

10

By the definition in frame 6, the association is a *function* from the hour of the day to the temperature since for each hour of the day there is associated exactly one temperature. The set whose elements are the hours of the day corresponds to set A in the definition. Thus, h, which represents an element of that set, is the independent variable. The temperature T is the dependent variable. This terminology is reasonable since we can *independently* select any hour, h, of the day and use the function to find the temperature, T, which *depends* on the hour selected.

If you feel you understand this now, go directly to 11. If you still feel uncertain, you should reread frames 3 to 10 and then go to 11.

11

Let's next consider how to specify a function. One way is to list in detail the association between the corresponding elements of the two sets. Another way is to present a rule for finding the dependent variable in terms of the independent variable. Often this rule is in the form of an equation. Thus, for example, a function associating the independent variable, t, with the dependent variable, S, could be specified by the equation

$$S = 2t^2 + 6t.$$

This equation defines the function since it associates exactly one value of the variable, S, with each value of the variable, t.

Strictly speaking, the function isn't completely specified until we give the allowed values (the domain) of the independent variable. There is a simple convention which we shall follow here: unless otherwise stated, the independent variable can be any real number for which the dependent variable is also a real number. Therefore, in the above example t can have any real value. On the other hand, if we are given a function defined by $y = \sqrt{x}$, then x is restricted to all the non-negative real numbers.

In most mathematics discussions both the independent and dependent variables are usually pure numbers, such as 5.1 or $\sqrt{7}$. However, in applications, the variables often have dimensions or units of measure, such as 5.1 seconds, $\sqrt{7}$ miles.

Go to 12.

12

    We usually represent a function by a letter such as f. If the independent variable is x, the dependent variable associated by the function f is often written as $f(x)$ and is read "f of x". The parentheses in $f(x)$ are part of the notation and indicate that the quantity enclosed by them represents the independent variable. Thus $f(x)$ does *not* mean f times x even though the parenthesis has that meaning in $5(3 + 2) = 25$. The ambiguous use of parenthesis may cause you a bit of confusion at first, but you should soon have little trouble telling when the parentheses are part of the symbol $f(x)$. One advantage of this notation is that the value of the dependent variable, say for $x = 3$, can be indicated by $f(3)$.

Go to 13.

13

    In mathematics the symbol x is most frequently used for an independent variable, f most often represents the function, and $y = f(x)$ usually denotes the dependent variable. However, any other symbols may be used for the function, the independent variable, and the dependent variable. For example, we might have $z = H(r)$ which is read as "z equals H of r". For the example in frame 11, $S = 2t^2 + 6t$, we could equally well have written

$$F(t) = 2\,t^2 + 6t,$$

in which case

$$S = F(t)$$

    Now that we know what a function means in the abstract, let's move along to a discussion of graphs.

Go to the next section,
frame 14.

Section 2. GRAPHS

14

    If you know how to plot graphs of functions, you should skip on to frame 19. Otherwise,

go to 15.

15

    A convenient way to represent a function defined by $y = f(x)$ is to plot a graph. Just for the record let us recall how to construct coordinate axes. · First, we construct a pair of mutually perpendicular intersecting lines and think of one of the lines as running vertically and the other as running horizontally. The horizontal line is usually called the *horizontal axis*, or *x*-axis, and the vertical line the *vertical axis*, or *y*-axis. The point of intersection is called the *origin*, and the axes together are called the *coordinate axes*.

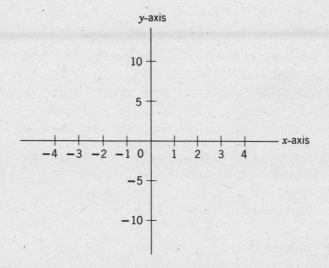

    Next we select a convenient unit of length and starting from the origin as zero, we mark off a number scale on the *x*-axis, positive to the right and negative to the left. In the same way we mark off a scale along the *y*-axis with positive numbers going upward and negative downward. The scale of the *y*-axis does not need to be the same as that for the *x*-axis. In fact, *y* and *x* can have different units, such as distance and time.

Go to 16.

16

We can represent one specific pair of values associated by the function in the following way: Let *a* represent some particular value for the independent variable *x*, and let *b* indicate the corresponding value of $y = f(x)$. Thus, $b = f(a)$. On the *x*-axis, we locate a point corresponding to the number *a* using our number scale. This location is shown in the figure by the point *A*. On the *y*-axis, we locate a point corresponding to the number *b*, shown in the figure by point *B*.

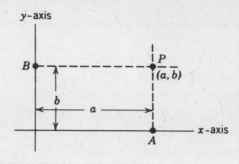

We now draw a line parallel to the *y*-axis through *A* and another line parallel to the *x*-axis through *B*. The point *P* at which these two lines intersect is taken to represent the pair of values (a, b) for *x* and *y* respectively.

The number *a* is often called the *x*-value of *P* or the *abscissa*, and the number *b* is called the *y*-value or *ordinate* of *P*. In the designation of a typical point by the notation (x, y) we will always list the abscissa, *x*, first in the parentheses before the comma and the ordinate, *y*, second, following the comma.

As a review of this terminology, encircle the correct answers below. For the point designated by (5, −3):

The abscissa is $\boxed{-5 \mid -3 \mid 3 \mid 5}$

The ordinate is $\boxed{-5 \mid -3 \mid 3 \mid 5}$

(Remember that the answers to all multiple choice questions are given at the bottom of the next page. Always check your answers before continuing.)

Go to 17.

17

The most direct way to plot the graph of a function $y = f(x)$ is to make a table of reasonably spaced values of x and of the corresponding values of $y = f(x)$. Then each pair of values (x,y) can be represented by a point as in the previous frame. A graph of the function is obtained by connecting the points with a smooth curve. Of course, the points on the curve may be only approximate. If we want an accurate plot we just have to be very careful and use many points. (On the other hand, crude plots are pretty good for most purposes.)

Go to 18.

18

As an example, here is a plot of the function $y = 3x^2$. A table of values of x and y is shown and these points are indicated on the graph.

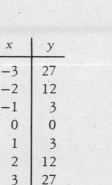

x	y
−3	27
−2	12
−1	3
0	0
1	3
2	12
3	27

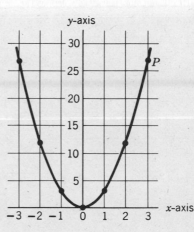

To test yourself, encircle below the pair of coordinates that corresponds to the point P indicated in the figure.

⌈(3,27) | (27,3) | none of these⌉

Check your answer. If correct, go on to 19. If incorrect study frame 16 once again and then go to 19.

Answers: (16) 5, −3

19

Here is a rather special function. Perhaps you have not come across it in quite this way yet. The function is called a *constant function* and associates a single fixed number, c, with all values of the independent variable, x. Hence, $f(x) = c$.

This is a peculiar function since the value of the dependent variable is the same for all values of the independent variable. Nevertheless, the relation $f(x) = c$ does associate exactly one value of $f(x)$ with each value of x as required in the definition of a function. It just so happens that all the values of $f(x)$ are the same.

Try to convince yourself that the graph of the constant function $y = f(x) = 3$ is a straight line parallel to the x-axis passing through the point $(0,3)$ as shown in the figure.

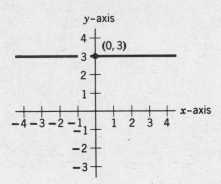

Go to 20.

Answer: (18) (3,27).

20

Another simple function is the *absolute value function*. The absolute value of x is indicated by the symbols $|x|$. The absolute value of a number, x, determines the size or magnitude of the number without regard to its sign. Therefore,

$$|-3| = |3| = 3.$$

Now we will define $|x|$ in a general way. But first we should recall the inequality symbols:

$a > b$ means a is greater than b.

$a \geq b$ means a is greater than or equal to b.

$a < b$ means a is less than b.

$a \leq b$ means a is less than or equal to b.

With this notation we can define the absolute value function, $|x|$, by the following two rules:

$$|x| = x \text{ if } x \geq 0$$
$$= -x \text{ if } x < 0.$$

Go to 21.

21

A good way to show the behavior of a function is to plot its graph. Therefore, as an exercise, plot a graph of the function $y = |x|$ in the accompanying figure.

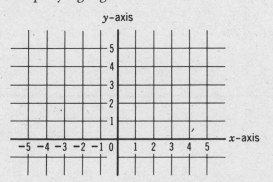

To check your answer,
go to 22.

22

The correct graph for $|x|$ is

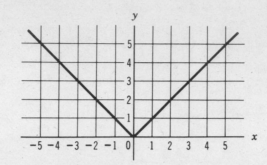

This can be seen by preparing a table of x and y values as follows:

x	$y = |x|$
−4	+4
−2	+2
0	0
+2	+2
+4	+4

These points may be plotted as in frames 16 and 18 and the lines drawn with the results in the above figure.

With this introduction on functions and graphs, we are now going to take a quick look at some elementary functions which are very important and with which you should become familiar.

These functions are: linear, quadratic, trigonometric, exponential, and logarithmic.

Proceed to Section 3,
frame 23.

Section 3. LINEAR AND QUADRATIC FUNCTIONS

23

A function defined by an equation in the form $y = mx + b$, where m and b are constants, is called a *linear* function because its plot is a straight line. This is a simple and useful function and you should really become familiar with it.

Here is an example: encircle the letter which identifies the graph of

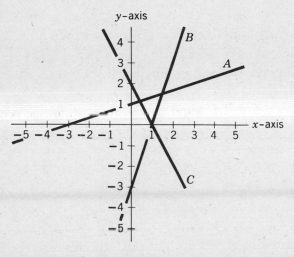

$$y = 3x - 3.$$

$$[A \mid B \mid C]$$

The correct answer is on the bottom of the next page. If you missed this or if you do not feel entirely sure of the answer, go to 24.

Otherwise, go to 25.

24

You were given the function $y = 3 x - 3$. The table below gives a few values of x and y.

A few of these points are shown on the graph, and a straight line has been drawn through them. This is line B of the figure in frame 23.

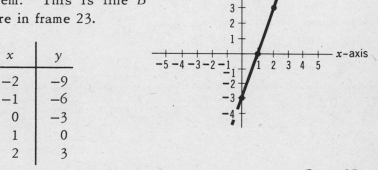

x	y
-2	-9
-1	-6
0	-3
1	0
2	3

Go to 25.

25

Here is the graph of a typical linear equation. Let us take any two points on the line, (x_2, y_2) and (x_1, y_1). We define the *slope* of the line in the following way:

$$\text{slope} = \frac{y_2 - y_1}{x_2 - x_1}.$$

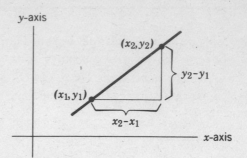

By multiplying the top and the bottom of this expression by -1, it should be apparent that the slope also equals $(y_1 - y_2)/(x_1 - x_2)$. The idea of slope will be very important in our later work, so let's spend a little time learning more about it.

Go to 26.

26

If the x and y scales are the same, as in the figure, then the slope is the ratio of *vertical* distance to *horizontal* distance as we go from one point on the line to another, providing we take the sign of each line segment as in the equation of frame 25. If the line is vertical, the slope is infinite (or, more strictly, undefined). It should be clear that the slope is the same for any pair of two separate points on the line.

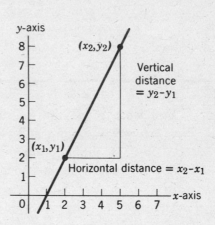

Go to 27.

Answer: (23) *B*

27

If the vertical and horizontal scales are not the same, the slope is still defined by

$$\text{slope} = \frac{\text{vertical distance}}{\text{horizontal distance}},$$

but now the distance is measured using the appropriate scale. For instance, the two figures below may look similar, but the slopes are quite different. In the first figure the x and y scales are identical, and the slope is 1/2. In the second figure the y scale has been changed by a factor of 100, and the slope is 50.

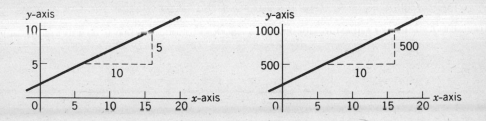

Since the slope is the ratio of two lengths, the slope is a pure number if the lengths are pure numbers. However, if the variables have different dimensions, the slope will also have dimensions.

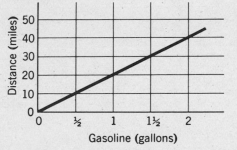

At the left is a plot of the distance travelled by a car vs. the amount of gasoline consumed. Here the slope has the unit of miles/ gallon (or miles per gallon). What is the slope of the line shown?

Slope = ⌐10 | 20 | 30 | 40⌐ miles/gallon

If right, go to 29.
Otherwise, go to 28.

28

To evaluate the slope, let us find the co-ordinates of two points on the line. For instance, A has the co-ordinates (2 gallons, 40 miles) and B has the co-ordinates (1/2 gallon, 10 miles). Therefore, the slope is

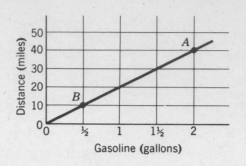

$$\frac{(40-10) \text{ miles}}{(2-1/2) \text{ gallons}} = \frac{30 \text{ miles}}{3/2 \text{ gallons}} = 20 \text{ miles per gallon.}$$

Of course, we would have obtained the same value for the slope no matter which two points we used, since the ratio of vertical distance to horizontal distance is the same everywhere.

Go to 29.

29

Here is another way to find the slope of a straight line if its equation is given. Our linear equation is in the form $y = mx + b$. The slope is given by

$$\text{slope} = \frac{y_2 - y_1}{x_2 - x_1}.$$

Substituting in the above expression for y, we have

$$\text{slope} = \frac{mx_2 + b - (mx_1 + b)}{x_2 - x_1} = \frac{mx_2 - mx_1}{x_2 - x_1} = \frac{m(x_2 - x_1)}{x_2 - x_1} = m.$$

What is the slope of $y = 7x - 5$?

$\boxed{5/7 \mid 7/5 \mid -5 \mid -7 \mid 5 \mid 7}$

If right, go to 31.
Otherwise, go to 30.

Answer: (27) 20 miles/gallon.

30

The equation $y = 7x - 5$ can be written in the standard form $y = mx + b$ if $m = 7$ and $b = -5$. Since slope $= m$, the line given has a slope of 7.

Go to 31.

31

The slope of a line can be positive (greater than 0), negative (less than 0), or 0.　An example of each is shown graphically below

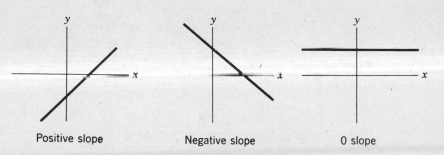

Positive slope　　　　Negative slope　　　　0 slope

Note how a line with positive slope rises in going from left to right, while a line with negative slope falls in going from left to right.

Indicate whether the slope of the graph of each of the following equations is positive, negative or 0 by encircling your choice.

Equation　　　　　Slope

1) $y = 2x - 5$　　　　$[+ \mid - \mid 0]$

2) $y = -3x$　　　　$[+ \mid - \mid 0]$

3) $p = q - 2$　　　　$[+ \mid - \mid 0]$

4) $y = 4$　　　　$[+ \mid - \mid 0]$

If all right, go to 33.
If you made any mistakes,
go to 32.

Answer: (29)　7

32

Here are the explanations to the questions in 31.

In frame 29 we saw that for the standard form of the linear equation, $y = mx + b$, the slope is equal to m.

1) $y = 2x - 5$. Here $m = 2$ and the slope is 2. Clearly this is a positive number. See Fig. 1 below.

2) $y = -3x$. Here $m = -3$. The slope is -3, which is negative. See Fig. 2 below.

3) $p = q - 2$. In this equation the variables are p and q, rather than y and x. Hence the standard form with these variables io $p - mq + h$ Here $m = 1$, which is positive. See Fig. 3.

4) $y = 4$. This is an example of a constant function. Here $m = 0$, $b = 4$, and the slope is 0. See Fig. 4.

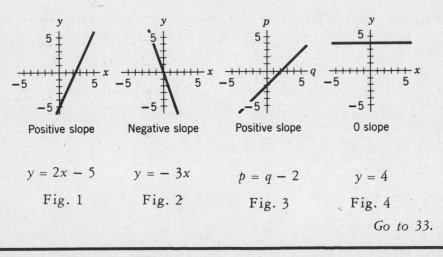

| Positive slope | Negative slope | Positive slope | 0 slope |

| $y = 2x - 5$ | $y = -3x$ | $p = q - 2$ | $y = 4$ |
| Fig. 1 | Fig. 2 | Fig. 3 | Fig. 4 |

Go to 33.

Answers: (31) +, −, +, 0

33

Here is an example of a linear equation in which the slope has a familiar meaning. The graph at the right shows the position, S, on a straight road of a car at different times. The position $S = 0$ means the car is at the starting point.

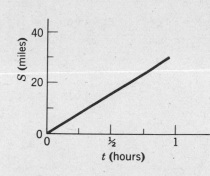

You should be able to guess which word to fill in the blank below.

The slope of the line has the same value as the car's _____.

To see the correct answer, go to 34.

34

The slope of the line has the same value as the car's *velocity* (or its *speed*).

The reason for this is that the slope is given by the ratio of the distance traveled to the time required. But, by definition, the velocity is also the distance traveled divided by the time. Thus the value of the slope of the line is equal to the velocity.

Go to 35.

35

Now let's look at another type of equation. An equation in the form $y = ax^2 + bx + c$, where a, b, and c are constants, is called a *quadratic* equation and its graph is called a *parabola*. Two typical parabolas are shown in the figure.

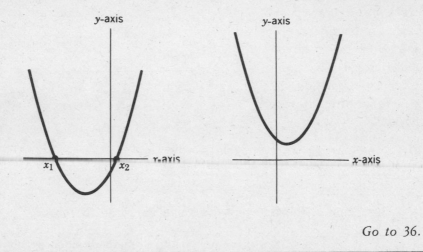

Go to 36.

36

The values of x at $y = 0$, shown by x_1 and x_2 in the left figure correspond to values of x which satisfy $ax^2 + bx + c = 0$ and are called the *roots* of the equation. Not all quadratic equations have real roots. For example, the curve on the right represents an equation with no real value of x at $y = 0$.

Although you will not need to find the roots of any quadratic equation later in this book, you may want to know the formula anyway. If you would like to see a discussion of this, go to frame 37. Otherwise, skip to frame 39.

37

The equation $ax^2 + bx + c = 0$ has two roots, and these are given by

$$x_1 = \frac{-b + \sqrt{b^2 - 4ac}}{2a} \qquad x_2 = \frac{-b - \sqrt{b^2 - 4ac}}{2a}$$

The subscripts 1 and 2 serve merely to identify the two roots. They can be omitted, and the above two equations can be summarized by

$$x = \frac{-b \pm \sqrt{b^2 - 4ac}}{2a}$$

We will not prove these results, though they can be checked by substituting the values for x in the original equation.

Here is a practice problem on finding roots: Which answer correctly gives the roots of $3x - 2x^2 = 1$?

a) $\frac{1}{4} (3 + \sqrt{17}); \frac{1}{4} (3 \quad \sqrt{17})$

b) $-1; -\frac{1}{2}$

c) $\frac{1}{4}; -\frac{1}{4}$

d) $1; \frac{1}{2}$

Encircle the letter of the correct answer.

$[a \mid b \mid c \mid d]$

If you got the right answer,
go to 39.
If you missed this, go to 38.

38

Here is the solution to the problem in frame 37.

The equation $3x - 2x^2 = 1$ can be written in the standard form

$$2x^2 - 3x + 1 = 0.$$

Here $a = 2$, $b = -3$, $c = 1$.

$$x = \frac{1}{2a}\left[-b \pm \sqrt{b^2 - 4ac}\right] = \frac{1}{4}\left[-(-3) \pm \sqrt{3^2 - 4 \times 2 \times 1}\right]$$

$$= \frac{1}{4}(3 \pm 1)$$

$$x_1 = \frac{1}{4} \times 4 = 1$$

$$x_2 = \frac{1}{4} \times 2 = \frac{1}{2}$$

                                                                Go to 39.

39

This ends our brief discussion of linear and quadratic functions. Perhaps you would like some more practice on these topics before continuing. If so, try working review problems 1–5 at the back of the book. In Chapter IV there is a concise summary of the material we have had so far, which you may find useful.

                                        Whenever you are ready,
                                        go to Section 4,
                                        frame 40.

Answer: (37) d

Section 4.   TRIGONOMETRY

40

Trigonometry involves angles, so here is a quick review of the units we use to measure angles. There are two important units: *degrees* and *radians*.

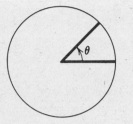

Degrees: A circle is divided into 360 equal angles. Each of these angles is 1° (one degree). (Each degree is further subdivided into 60 minutes [60'], and each minute is subdivided into 60 seconds [60"]. However, we will not need to use such fine divisions here.) It follows from this that a semi-circle contains 180°. Which of the following angles is equal to the angle θ (Greek letter theta) shown in the figure?

[25° | 45° | 90° | 180°]

If right, go to 42.
Otherwise, go to 41.

41

To find the angle θ, let's first look at a related example.

The angle shown is a right angle. Since there are four right angles in a full circle, it is apparent that the angle equals

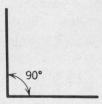

$$\frac{360°}{4} = 90°.$$

The angle θ shown in frame 40 is just half as big as the right angle; thus it is 45°.

Here is a circle divided into equal segments by three straight lines. Which angle equals 240°?

[a | b | c]

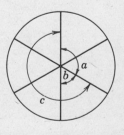

Go to 42.

42

The second unit of angular measure, and the most useful for our later work, is the *radian*.

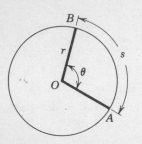

To find the value of an angle in radians, we draw a circle of radius r, about the vertex, 0, of the angle so that it intersects the sides of the angle at two points, shown in the figure as A and B. The length of the arc between A and B is designated by s. Then,

$$\theta \text{ (in radians)} = \frac{s}{r} = \frac{\text{length of arc}}{\text{radius}}$$

To see whether you have caught on, answer this question: there are 360 degrees in a circle; how many radians are there?

$$\boxed{1 \mid 2 \mid \pi \mid 2\pi \mid 360/\pi}$$

If right, go to 44.
Otherwise, go to 43.

43

The circumference of a circle is given by $c = 2\pi r$. The length of an arc going completely around a circle is then $2\pi r$, and the angle enclosed is $2\pi r/r = 2\pi$ radians, as shown in the figure on the left. In the figure on the right the angle θ subtends an arc $s = r$. Encircle the answer which gives θ.

$$\boxed{1 \text{ rad} \mid 1/4 \text{ rad} \mid 1/2 \text{ rad} \mid \pi \text{ rad} \mid \text{none of these}}$$

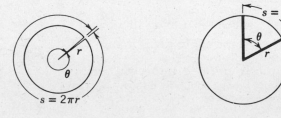

Go to 44.

Answers: (40) 45°;   (41) c

44

Because many of the relations we develop later are much simpler when the angles are measured in radians, we will stick to the rule that *all angles will be in radians unless they are marked in degrees.*

Sometimes the word radian is written out in full, sometimes it is abbreviated to rad, but usually it is omitted entirely. Thus: $\theta = 0.6$ means 0.6 radian; $27°$ means 27 degrees; $\pi/3$ rad means $\pi/3$ radians.

Go on to 45.

45

Since 2π rad $= 360°$, the formula for converting from degrees to radians is

$$1 \text{ rad} = \frac{360°}{2\pi}$$

and

$$1 \text{ degree} = \frac{2\pi \text{ rad}}{360}$$

Now you should be able to do the following problems (encircle the correct answer):

$$60° = \boxed{2\ \pi/3 \mid \pi/3 \mid \pi/4 \mid \pi/6}\ \text{rad}$$

$$\pi/8 = \boxed{22\ 1/2° \mid 45° \mid 60° \mid 90°}$$

Which angle is closest to 1 rad? (Remember that $\pi = 3.14 \dots$)

$$\boxed{30° \mid 45° \mid 60° \mid 90°}$$

If right, go to 47.
If you made any mistake,
go to 46.

Answers: (42) 2π
         (43) 1 rad

46

Here are the solutions to the problems in frame 45. From the formulas in frame 45, it should be easy to see that

$$60° = 60 \times \frac{2\pi \text{rad}}{360} = \frac{2\pi \text{rad}}{6} = \frac{\pi}{3} \text{rad}$$

$$\frac{\pi}{8} \text{rad} = \frac{\pi}{8} \times \frac{360°}{2\pi} = \frac{360°}{16} = 22\frac{1}{2}°.$$

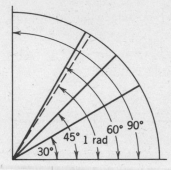

$1 \text{rad} = \frac{360°}{2\pi}$. Since 2π is just a little greater than 6, 1 rad is slightly less than $\frac{360°}{6} = 60°$. (Actually, 1 rad = 57° 18′.) The figure shows all the angles in this question.

Go to 47.

47

In the circle shown, CG is perpendicular to AE.

arc AB = arc BC = arc AH.
arc AD = arc DF = arc FA.
(arc AB means the length of the arc along the circle between A and B, going the shortest way)

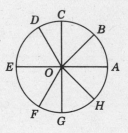

We will designate an angle by three letters as in the following example: $\angle\ AOB$ (read as "angle AOB").

You should be able to work out the following:

$\angle AOD = \boxed{60° \mid 90° \mid 120° \mid 150° \mid 180°}$

$\angle FOH = \boxed{15° \mid 30° \mid 45° \mid 60° \mid 75° \mid 90°}$

$\angle BOH = \boxed{1/4 \mid 1 \mid \pi/2 \mid \pi/4 \mid \pi/8}$

If you did all these correctly, go to 49.
If you made any mistakes, go to 48.

Answers: (45) $\pi/3$, 22½°, 60°

48

    By studying the figure in frame 47, and the definitions in frames 40 and 42, you should be able to track down your error in answering 47. After doing this, try the following problems.

$90° = \boxed{2\pi \mid \pi/6 \mid \pi/2 \mid \pi/8 \mid 1/4}$

$3\pi = \boxed{240° \mid 360° \mid 540° \mid 720°}$

$\pi/6 = \boxed{15° \mid 30° \mid 45° \mid 60° \mid 90° \mid 120°}$

Go to 49.

49

    Our next task is to review the trigonometric functions. One use of these functions is to relate the sides of triangles, particularly right triangles, to their angles. We will get to this application shortly. However, right now we will define the trigonometric functions in a more general and more useful way.

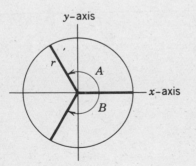

Here is a circle of radius r drawn with x and y axes, as shown. We will choose the positive x-axis as a reference line and, for the purpose of this section, we will measure angles from the reference line. An angle formed by rotating in a counterclockwise direction is positive; an angle formed by moving in a clockwise direction is negative. As an example, the angle A is positive and B is negative, as shown in the figure.

Go to 50.

Answers: (47) 120°, 75°, $\pi/2$

50

Do you remember the general definitions of the trigonometric functions of an angle θ? If you do, test yourself with the quiz below. If you don't, go right on to frame 51.

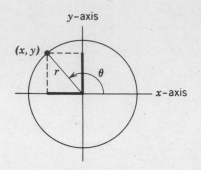

The trigonometric functions of θ can be expressed in terms of the co-ordinates x and y and the radius of the circle, $r = \sqrt{x^2 + y^2}$. These are shown in the figure. Try to fill in the blanks (the answers are in frame 51):

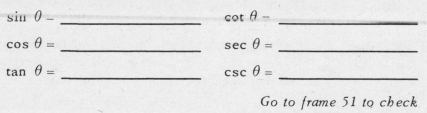

sin θ = _____   cot θ = _____

cos θ = _____   sec θ = _____

tan θ = _____   csc θ = _____

*Go to frame 51 to check
your answers.*

Answers: (48) $\pi/2$, 540°, 30°

51

Here are the definitions of the trigonometric functions:

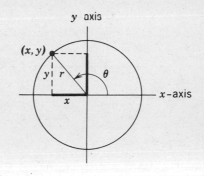

sine: $\sin \theta = \dfrac{y}{r}$

cosine: $\cos \theta = \dfrac{x}{r}$

tangent: $\tan \theta = \dfrac{y}{x}$

cotangent: $\cot \theta = 1/\tan \theta = \dfrac{x}{y}$

secant: $\sec \theta = 1/\cos \theta = \dfrac{r}{x}$

cosecant: $\csc \theta = 1/\sin \theta = \dfrac{r}{y}$

For the angle shown in the figure, x is negative and y is positive ($r = \sqrt{x^2 + y^2}$ and is always positive) so that $\cos \theta$, $\tan \theta$, $\cot \theta$, and $\sec \theta$ are negative.

After you have studied
these, go to 52.

52

Here is a circle with a radius of 5. The point shown is $(-3, -4)$. On the basis of the definition in the last frame, you should be able to answer the following:

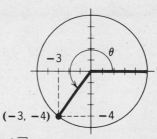

$\sin \theta =$ $\lceil 3/5 \mid 5/3 \mid 3/4 \mid -4/5 \mid -3/5 \mid 4/3 \rceil$

$\cos \theta =$ $\lceil 3/5 \mid 5/3 \mid 3/4 \mid -4/5 \mid -3/5 \mid 4/3 \rceil$

$\tan \theta =$ $\lceil 3/5 \mid 5/3 \mid 3/4 \mid -4/5 \mid -3/5 \mid 4/3 \rceil$

If all right, go to 55.
Otherwise, go to 53.

53

Perhaps you had difficulty because you did not realize that x and y have different signs in different quadrants (quarters of the circle) while r, a radius, is always positive. Try this problem.

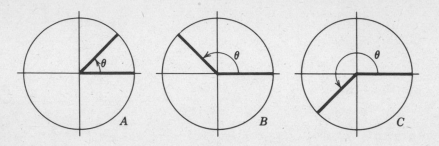

A   B   C

Indicate whether the function required is positive or negative, for each of the figures, by checking the correct box.

	Fig. A + −	Fig. B + −	Fig. C + −
sin θ			
cos θ			
tan θ			

See frame 54 for the correct answers.

54

Here are the answers to the questions in frame 53.

	Fig. A + −	Fig. B + −	Fig. C + −
sin θ	✓	✓	✓
cos θ	✓	✓	✓
tan θ	✓	✓	✓

Go to 55.

Answers: (52) $-4/5$, $-3/5$, $4/3$

55

In the figure both θ and $-\theta$ are shown. The trigonometric functions for these two angles are simply related. Can you do these problems? Encircle the correct sign.

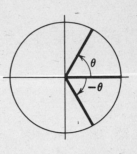

$\sin(-\theta) = \boxed{+ \mid -}\ \sin\theta$

$\cos(-\theta) = \boxed{+ \mid -}\ \cos\theta$

$\tan(-\theta) = \boxed{+ \mid -}\ \tan\theta$

Go to 56.

56

There are many relationships among the trigonometric functions. For instance, using $x^2 + y^2 = r^2$, we have

$$\sin^2\theta = \frac{y^2}{r^2} = \frac{r^2 - x^2}{r^2} = 1 - \left(\frac{x}{r}\right)^2 = 1 - \cos^2\theta$$

Try these:

1) $\sin^2\theta + \cos^2\theta = \boxed{\sec^2\theta \mid 1 \mid \tan^2\theta \mid \cot^2\theta}$

2) $1 + \tan^2\theta = \boxed{1 \mid \tan^2\theta \mid \cot^2\theta \mid \sec^2\theta}$

3) $\sin^2\theta - \cos^2\theta = \boxed{1 - 2\cos^2\theta \mid 1 - 2\sin^2\theta \mid \cot^2\theta \mid 1}$

If any mistakes, go to 57.
Otherwise, go to 58.

57

Here are the solutions to problem 16.

(1) $\sin^2\theta + \cos^2\theta = \dfrac{y^2}{r^2} + \dfrac{x^2}{r^2} = \dfrac{x^2 + y^2}{r^2} = \dfrac{r^2}{r^2} = 1$

This is an important identity which you should remember.

(2) $1 + \tan^2\theta = 1 + \dfrac{\sin^2\theta}{\cos^2\theta} = \dfrac{\cos^2\theta + \sin^2\theta}{\cos^2\theta} = \dfrac{1}{\cos^2\theta} = \sec^2\theta.$

(3) $\sin^2\theta - \cos^2\theta = 1 - \cos^2\theta - \cos^2\theta = 1 - 2\cos^2\theta = 2\sin^2\theta - 1.$

Go to 58.

58

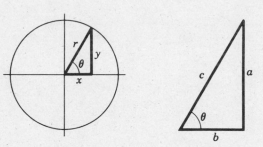

The trigonometric functions are particularly useful when applied to right triangles (triangles with one right angle). In this case θ is always acute (less than 90° or $\pi/2$). You should be able to write the trigonometric functions in terms of the sides a, b and c of the triangle shown. Fill in the blanks.

sin θ = _____ .   cot θ = _____

cos θ = _____   sec θ = _____

tan θ = _____   csc θ = _____

Check your answer in 59.

Answers: (55) −, +, −;   (56) 1, $\sec^2\theta$, $1 - 2\cos^2\theta$

59

The answers are:

$$\sin\ \theta = \frac{a}{c} = \frac{\text{opposite side}}{\text{hypotenuse}}$$

$$\cos\ \theta = \frac{b}{c} = \frac{\text{adjacent side}}{\text{hypotenuse}}$$

$$\tan\ \theta = \frac{a}{b} = \frac{\text{opposite side}}{\text{adjacent side}}$$

$$\cot\ \theta = \frac{b}{a} = \frac{\text{adjacent side}}{\text{opposite side}}$$

$$\sec\ \theta = \frac{c}{b} = \frac{\text{hypotenuse}}{\text{adjacent side}}$$

$$\csc\ \theta = \frac{c}{a} = \frac{\text{hypotenuse}}{\text{opposite side}}$$

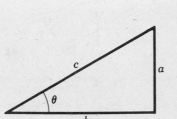

These results follow from the definitions in Frame 51, providing we let a, b, and c correspond to y, x, and r respectively. (Remember that here θ is less than 90°.) If you are not familiar with the terms opposite side, adjacent side and hypotenuse, they should be evident from the figure.

Go to 60.

60

The following problems refer to the figure shown (ϕ is the Greek letter "phi").

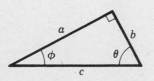

$$\cos\ \theta = \boxed{b/c \mid a/c \mid c/a \mid c/b \mid b/a \mid a/b}$$

$$\tan\ \phi = \boxed{b/c \mid a/c \mid c/a \mid c/b \mid b/a \mid a/b}$$

If all right, go to 62.
Otherwise, go to 61.

61

You may have become confused because the triangle was drawn in a new position. Review the definitions in 51, and then do the problems below:

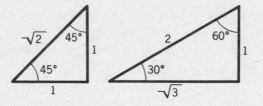

$\sin\ \theta = \boxed{l/n \mid n/l \mid m/n \mid m/l \mid n/m \mid l/m}$

$\tan\ \phi = \boxed{l/n \mid n/l \mid m/n \mid m/l \mid n/m \mid l/m}$

If you missed either of these you will have to put in more work learning the definitions. Unfortunately, with definitions there is no substitute for buckling down to work and memorizing them.

Meanwhile go to 62.

62

You should be familiar with the 45° and the 30° and 60° right triangles whose sides are proportional to the numbers shown.

Try the problems below:

$\cos 45° = \boxed{1/2 \mid 1/\sqrt{2} \mid 2\sqrt{2} \mid 2}$

$\sin 30° = \boxed{3 \mid \sqrt{3}/2 \mid 2/3 \mid 1/2}$

$\sin 45° = \boxed{1/2 \mid 1/\sqrt{2} \mid 2\sqrt{2} \mid 2}$

$\tan 30° = \boxed{1 \mid \sqrt{3} \mid 1/\sqrt{3} \mid 2}$

Make sure you understand
these problems.
Then go to 63.

Answers: (60) *b/c, b/a*

63

Because the angle $2\pi + \theta$ is equivalent to θ as far as the diagram is concerned, we can add 2π to any angle without changing the value of the trigonometric functions. Since the sine and cosine functions repeat their values whenever θ increases by 2π, we say that the functions are *periodic* in θ with a period of 2π. Tan θ and cot θ are also periodic, but they have a period of π.

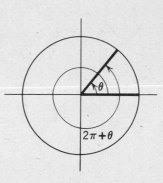

Go to 64.

64

You should become familiar with the graphs of the trigonometric functions. For instance, here is a graph of sin θ.

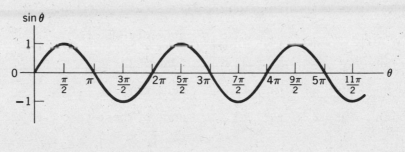

Go to 65.

Answers: (61) *l/n, m/l*
          (62) $1/\sqrt{2}, 1/2, 1/\sqrt{2}, 1/\sqrt{3}$

65

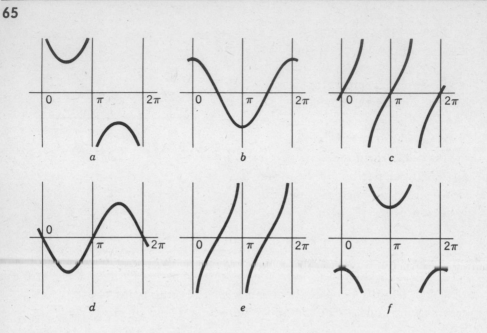

a

b

c

d

e

f

Try to figure out which graph represents each function.

cos θ: $[\,a\mid b\mid c\mid d\mid e\mid f\mid$ none of these$\,]$

tan θ: $[\,a\mid b\mid c\mid d\mid e\mid f\mid$ none of these$\,]$

sin $(-\theta)$: $[\,a\mid b\mid c\mid d\mid e\mid f\mid$ none of these$\,]$

tan $(-\theta)$: $[\,a\mid b\mid c\mid d\mid e\mid f\mid$ none of these$\,]$

If you got these all right,
go to 67.
Otherwise go to 66.

66

Knowing the values of the trigonometric functions at a few important points will help you identify them. Try these (∞ is the symbol for infinity):

sin $(0°) = [\,0\mid 1\mid -1\mid -\infty\mid +\infty\,]$

cos $(90°) = [\,0\mid 1\mid -1\mid -\infty\mid +\infty\,]$

tan $(45°) = [\,0\mid 1\mid -1\mid -\infty\mid +\infty\,]$

Go to 67.

67

It is very handy to know the sine and cosine of the sum and the difference of two angles

Do you remember the formulas from previous studies of trigonometry? If not, go to 68. If you do, try the quiz below.

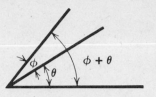

sin $(\theta + \phi)$ = _____

cos $(\theta + \phi)$ = _____

*Go to frame 68 to see
the correct answer.*

68

Here are the required formulas. They are derived in Appendix A1.

$$\sin (\theta + \phi) = \sin \theta \cos \phi + \cos \theta \sin \phi$$
$$\cos (\theta + \phi) = \cos \theta \cos \phi - \sin \theta \sin \phi$$

(Note that tan $(\theta + \phi)$ and cot $(\theta + \phi)$ can be obtained from these formulas and the relation tan θ = sin θ/cos θ.)

By using what you have already learned, try to circle the correct sign in each of the following:

(a) sin $(\theta - \phi)$ = $\boxed{+\ |\ -}$ sin θ cos ϕ $\boxed{+\ |\ -}$ cos θ sin ϕ

(b) cos $(\theta - \phi)$ = $\boxed{+\ |\ -}$ cos θ cos ϕ $\boxed{+\ |\ -}$ sin θ sin ϕ

*If right, go to 70.
If wrong, go to 69.*

Answers: (65) *b, c, d,* none of these
               (66) 0, 0, 1

69

If you made a mistake in problem 68 you should recall from frame 55 that $\sin(-\phi) = -\sin(\phi)$
$$\cos(-\phi) = +\cos(\phi)$$

Then

$$\sin(\theta - \phi) = \sin(\theta)\cos(-\phi) + \cos(\theta)\sin(-\phi)$$
$$= \sin(\theta)\cos(\phi) - \cos(\theta)\sin(\phi)$$

$$\cos(\theta - \phi) = \cos(\theta)\cos(-\phi) - \sin(\theta)\sin(-\phi)$$
$$= \cos(\theta)\cos(\phi) + \sin(\theta)\sin(\phi)$$

Go to 70.

70

By using the expressions for $\sin(\theta + \psi)$ and $\cos(\theta + \phi)$ one can obtain the formulas for $\sin 2\theta$ and $\cos 2\theta$. Simply let $\theta = \phi$. Fill in the blanks.

$$\sin 2\theta = \underline{\hspace{4cm}}$$

$$\cos 2\theta = \underline{\hspace{4cm}}$$

See 71 for the correct answers.

71

$$\sin 2\theta = 2\sin\theta\cos\theta$$

$$\cos 2\theta = \cos^2\theta - \sin^2\theta$$

Go to 72.

Answers: (68a) +, −;   (68b) +, +

72

It is often convenient to use the *inverse trigonometric function*, which designates the angle for which the trigonometric function has the specified value. Thus the inverse trigonometric function to $y = \sin \theta$ is $\theta = \arcsin y$, which is read as the "arc-sine of y" and stands for the angle whose sine is y. The arcos y, arctan y, etc., are similarily defined. Try this problem.

$$\arcsin (1/2) = \underline{\hspace{4cm}}$$

*See frame 73 for the
right answer.*

73

Arcsin $(1/2) = 30°$ since sin $(30°) = 1/2$ (see frame 62) and arcsin y is defined as the angle whose sine is y.

Now go on to the next section, which is the last one in our review.

Go to frame 74.

74

Do you already understand exponentials?   If not, go to 75. If you do, try the short quiz below.

$$a^5 = \boxed{5^a \mid 5 \log a \mid a \log 5 \mid \text{none of these}}$$

$$a^{b+c} = \boxed{a^b \times a^c \mid a^b + a^c \mid ca^b \mid (b+c) \log a}$$

$$a^f / a^g = \boxed{(f-g) \log a \mid a^{f/g} \mid a^{(f-g)} \mid \text{none of these}}$$

$$a^0 = \boxed{0 \mid 1 \mid a \mid \text{none of these}}$$

$$(a^b)^c = \boxed{a^b \times a^c \mid a^{(b+c)} \mid a^{(bc)} \mid \text{none of these}}$$

If any mistakes, go to 75.
Otherwise go to 76.

75

By definition a^m is the product of m factors of a.   Hence,

$$2^3 = 2 \times 2 \times 2 = 8, \text{ and } 10^2 = 10 \times 10 = 100.$$

Furthermore, by definition $a^{-m} = 1/a^m$.

It is easy to see, then, that

$$a^m \times a^n = a^{(m+n)}$$

$$a^m / a^n = a^{(m-n)}$$

$$a^0 = a^m / a^m = 1 \quad (m \text{ can be any integer})$$

$$(a^m)^n = a^{(mn)}$$

$$(ab)^m = a^m b^m$$

Go to 76.

76

Here are a few problems to do:

$3^2 = \boxed{6 \mid 8 \mid 9 \mid \text{none of these}}$

$1^3 = \boxed{1 \mid 3 \mid 1/3 \mid \text{none of these}}$

$2^{-3} = \boxed{-6 \mid 1/8 \mid -9 \mid \text{none of these}}$

$4^3/4^5 = \boxed{4^8 \mid 4^{-8} \mid 16^{-1} \mid \text{none of these}}$

> *If you did these all*
> *correctly, go to 78.*
> *If you made any mistakes,*
> *go to 77.*

77

Below are the solutions to problem 76. You should refer back to the rules in 75 if you don't understand them.

$3^2 = 3 \times 3 = 9$

$1^3 = 1 \times 1 \times 1 = 1$ ($1^m = 1$ for any m)

$2^{-3} = 1/2^3 = 1/8$

$4^3/4^5 = 4^{(3-5)} = 4^{-2} = 1/16 = 16^{-1}$

Now try these:

$(3^{-3})^3 = \boxed{1 \mid 3^{-9} \mid 3^{-27} \mid \text{none of these}}$

$5^2/3^2 = \boxed{(5/3)^2 \mid (5/3)^{-1} \mid 5^{-6} \mid \text{none of these}}$

$4^3 = \boxed{12 \mid 16 \mid 2^6 \mid \text{none of these}}$

Check your answers and try to track down any mistakes. Then go to 78.

Answers: (74) none of these, $a^b \times a^c$, $a^{(f-g)}$, 1, $a^{(bc)}$

78

Here are a few more problems.

$$10^0 = \boxed{0 \mid 1 \mid 10}$$

$$10^{-1} = \boxed{-1 \mid 1 \mid 0.1}$$

$$.00003 = \boxed{1/3 \times 10^{-3} \mid 10^{-3} \mid 3 \times 10^{-5}}$$

$$0.4 \times 10^{-4} = \boxed{4 \times 10^{-5} \mid 4 \times 10^{-3} \mid 2.5 \times 10^{-5}}$$

$$\frac{3 \times 10^{-7}}{6 \times 10^{-3}} = \boxed{1/2 \times 10^{10} \mid 5 \times 10^4 \mid 0.5 \times 10^{-4}}$$

If these were all correct,
go to 80.
If you made any mistakes,
go to 79.

79

Here are the solutions to problems 78:

$$10^0 = 1 \ (x^0 = 1, \text{ for any number except } 0)$$

$$10^{-1} = 1/10 = 0.1$$

$$.00003 = .00001 \times 3 = 3 \times 10^{-5}$$

$$0.4 \times 10^{-4} = (4 \times 10^{-1}) \times 10^{-4} = 4 \times 10^{-5}$$

$$\frac{3 \times 10^{-7}}{6 \times 10^{-3}} = \frac{3}{6} \times \frac{10^{-7}}{10^{-3}} = \frac{1}{2} \times 10^{(-7+3)} = 0.5 \times 10^{-4}$$

Go to 80.

Answers: (76) 9, 1, 1/8, 16^{-1}
        (77) 3^{-9}, $(5/3)^2$, 2^6

80

Let's briefly review fractional exponents. If $b^n = a$, then b is called the n'th *root* of a and is written $b = a^{1/n}$. Hence $16^{1/4}$ = (4'th root of 16) = 2. That is, $2^4 = 16$.

If $y = a^{m/n}$, where m and n are integers, then $y = [a^{1/n}]^m$. For instance:

$$8^{2/3} = (8^{1/3})^2 = 2^2 = 4.$$

Try this problem:

$$27^{-2/3} = \boxed{1/18 \mid 1/81 \mid 1/9 \mid -18 \mid \text{none of these}}$$

If right, go to 82.
If wrong, go to 81.

81

$$27^{-2/3} = [27^{1/3}]^{-2} = 3^{-2} = \frac{1}{9}$$

To check this, note that $\left[\frac{1}{9}\right]^{-3/2} = \left[\frac{1}{3}\right]^{-3} = 27.$

Here is another problem:

$$16^{3/4} = \boxed{12 \mid 8 \mid 6 \mid 64}$$

Go to 82.

82

Do these problems:

$$25^{3/2} = \boxed{125 \mid 5 \mid 15 \mid \text{none of these}}$$

$$(.00001)^{-3/5} = \boxed{.001 \mid 1000 \mid 10^{-15}/10^{-25}}$$

$$8^{-4/3} = \boxed{1/6 \mid 16 \mid 1/8 \mid 1/16}$$

If all your answers were correct, go to 84.
Otherwise go to 83.

Answer: (78) 1, 0.1, 3×10^{-5}, 4×10^{-5}, 0.5×10^{-4}

83

Here are the solutions to the problems in 82:

$$25^{3/2} = (25^{1/2})^3 = 5^3 = 125$$

$$[.00001]^{-3/5} = [10^{-5}]^{-3/5} = 10^{15/5} = 10^3 = 1000$$

$$8^{-4/3} = [8^{-1/3}]^4 = (1/2)^4 = 1/16$$

Here are a few more problems. Encircle the correct answers.

$$\left[\frac{27}{64} \times 10^{-6}\right]^{1/3} = \left[3/400 \;\middle|\; \frac{3}{16} \times 10^{-2} \;\middle|\; \frac{9}{64} \times 10^{-4}\right]$$

$$[49 \times 10^{-4}]^{1/4} = \left[\sqrt{7}/10 \;\middle|\; (10 \times 7)^{2} \;\middle|\; \sqrt{7/1000}\right]$$

Go to 84 after checking your answers.

84

Although our original definition of a^m only applied to integral values of m, we have also defined $(a^m)^{1/n} = a^{m/n}$ where both m and n are integers. Thus we have a meaning for a^p where p is either an integer or a fraction (ratio of integers). As yet we do not know how to evaluate a^p if p is an irrational number, as for instance π or $\sqrt{2}$. However, we can get around this problem in the following way: We can approximate an irrational number as closely as we desire by a fraction. For instance, π is approximately $314159/100000$. But this number is already in the form m/n, where m and n are integers, and we know how to evaluate $a^{m/n}$. Therefore, $y = a^x$, where x is any real number, is a meaningful expression in the sense that we can evaluate it to as much accuracy as we please.

To see if you understand this, try the following problem:

$$a^\pi a^x / a^3 = \left[a^{\pi x/3} \;\middle|\; a^{\pi + x - 3} \;\middle|\; a^{3\pi x} \;\middle|\; a^{(\pi + x)/3}\right]$$

If right, go to 86.
If wrong, go to 85.

Answers: (80) 1/9;   (81) 8;   (82) 125, 1000, 1/16

85

The rules given in frame 75 apply here as if all exponents were integers.

Hence $a^\pi a^x / a^3 = a^{\pi + x - 3}$

Here is another problem:

$\pi^2 \times 2^\pi = \boxed{1 \mid (2\pi)^{2\pi} \mid 2\pi^{(2+\pi)} \mid \text{none of these}}$

If right, go to 87.
If wrong, go to 86.

86

$\pi^2 \times 2^\pi$ is the product of two different numbers to two different exponents. None of our rules apply to this and, in fact, there is no way to simplify this expression.

Now go to 87.

87

If you do not clearly remember logarithms, go to 88. If you do, try the following test.

Let x be any positive number, and let $\log x$ represent the log of x to the base 10.

Then:

$10^{\log x} = $ _____

Go to 88 for the
correct answer.

Answers: (83) $3/400,\ \sqrt{7}/10$
(84) $a^{\pi + x - 3}$

88

The answer to 87 is x; in fact the logarithm of x to the base 10 is defined by.

$$10^{\log x} = x$$

That is, the logarithm of a number x is the power to which 10 must be raised to produce the number x itself. This definition only applies for $x > 0$. Here are two examples:

$100 = 10^2$, so $\log (100) = 2$

$.001 = 10^{-3}$, so $\log (.001) = -3$

Now try these problems:

$\log (1{,}000{,}000) = \boxed{1{,}000{,}000 \mid 6 \mid 60 \mid 600}$

$\log (1) = \boxed{0 \mid 1 \mid 10 \mid 100}$

If right, go to 90.
If wrong, go to 89.

89

$\log (1{,}000{,}000) = \log 10^6 = 6$
    (check, $10^6 = 1{,}000{,}000$)

$\log 1 = \log 10^0 = 0$
    (check, $10^0 = 1$)

You should be able to the following problems:

$\log (10^4/10^{-3}) = \boxed{10^7 \mid 1 \mid 10 \mid 7 \mid 70}$

$\log (10^n) \qquad = \boxed{10n \mid n \mid 10^n \mid 10/n}$

$\log (10^{-n}) \qquad = \boxed{-10n \mid -n \mid -10^n \mid -10/n}$

If you had trouble with these you should review the material in this section. Make sure you understand these problems, and then go to 90.

Answers: (85) none of these

90

See if you remember how to manipulate logarithms by trying the following problems, where a and b are any positive numbers:

$$\log (ab) \;=\; \big[\, \log a \times \log b \mid \log a + \log b \mid a \times \log b \,\big]$$

$$\log (a/b) = \big[\, \log a/\log b \mid - b \times \log a \mid \log a - \log b \,\big]$$

$$\log (a^n) \;=\; \big[\, n \times \log a \mid [\log a]^n \mid [\log a] + n \,\big]$$

If right, go to 92.
If wrong, go to 91.

91

We can derive the required rules by using the definition of $\log x$ together with the properties of exponentials. Recall that

$$a = 10^{\log a}, \;\; b = 10^{\log b}.$$

Hence

$$ab = 10^{\log a} \times 10^{\log b} = 10^{\left[\log a + \log b\right]}.$$

Taking the log of both sides and using the fact that $\log (10^x) = x$ then gives us

$$\log (ab) = \log 10^{\log a + \log b} = \log a + \log b.$$

Similarly,

$$a/b = 10^{\log a}\, 10^{-\log b} = 10^{\log a - \log b}.$$

so $\log (a/b) = \log a - \log b.$

Likewise; $a^n = [10^{\log a}]^n = 10^{n \log a}$

and $\log (a^n) = n \times \log a.$

Go to 92.

Answers: (88) 6, 0; (89) 7, n, $-n$

92

Try these problems:

if $\log\ n = -3$, $n = \boxed{1/3 \mid 1/300 \mid 1/1000}$

$10^{\log\ 100} = \boxed{10^{10} \mid 20 \mid 100 \mid \text{none of these}}$

$\dfrac{\log\ 1000}{\log\ 100} = \boxed{3/2 \mid 1 \mid -1 \mid 10}$

If right, go to 94.
If wrong, go to 93.

93

$10^{\log\ n} = n$, so if $\log n = -3$, $n = 10^{-3} = 1/1000$.

For the same reason,

$10^{\log\ 100} = 100$.

$\dfrac{\log\ 1000}{\log\ 100} = \dfrac{\log\ (10^3)}{\log\ (10^2)} = \dfrac{3}{2}$.

Try these problems:

$\dfrac{1}{2} \log\ (16) \quad = \boxed{2 \mid 4 \mid 8 \mid \log 2 \mid \log 4}$

$\log\ [\log\ (10)] = \boxed{10 \mid 1 \mid 0 \mid -1 \mid -10}$

Go to 94.

94

Here is a tricky problem — a good way to solve this one is by guesswork:

if $1 + \log\ (n) = n$, then $n = \boxed{0 \mid 1 \mid 10}$

If you did not get this, confirm that the given answer satisfies the equation.

Go on to 95.

Answers: (90) $\log a + \log b$, $\log a - \log b$, $n \times \log a$

95

In this section we have only discussed logs to the base 10. However, any positive number except 1 can be used as a base. Bases other than 10 are usually indicated by a subscript. For instance, the log of 8 to the base 2 is written $\log_2 8$. This log has the value of 3 since $2^3 = 8$. If our base is denoted by r, then the defining equation for $\log_r x$ is

$$\boxed{r^{\log_r x} = x}$$

All the relations explained in frame 91 of this section are true for logarithms *to any base* (provided, of course, that the same base is used for all the logarithms in each equation).

Go to 96.

96

This concludes our review. In order to do actual computations involving trigonometric functions and logarithms you will need their numerical values. You can find these in tables in many different books. For instance, *The Handbook of Chemistry and Physics* (Chemical Rubber Publishing Co.) contains convenient tables along with clear instructions on how to use them.

Before going on, there are a few features of this book you ought to know about. The last chapter, IV, consists of a summary of the first three chapters so that you can review what you have learned without going through all the problems. Take a look at that chapter before going on if you feel the need. In addition, starting on page 275 there is a list of review problems and answers arranged by section. These problems are for your use in case you want more practice.

As soon as you are ready,
go to Chapter II.

Answers: (92) 1/1000, 100, 3/2
         (93) log 4, 0
         (94) 1

CHAPTER II

DIFFERENTIAL CALCULUS

In this chapter you will learn

(1) what is meant by the limit of a function,

(2) how the derivative of a function is defined,

(3) how to interpret the derivative graphically,

(4) some shortcuts for finding derivatives,

(5) how to recognize the derivatives of the most common functions,

(6) how to find the maximum or minimum values of functions and how to apply differential calculus to problems involving position and velocity.

Section 1.   LIMITS

97

    Before tackling differential calculus, we must spend a little time learning about limits. The idea of a limit may very well be new to you, but it is at the heart of calculus and you should be particularly careful to understand the material in this section before going on. Once you have a real feeling for what is meant by a limit you will be able to grasp the ideas of differential calculus quite readily.

    Limits are so important in calculus that we will discuss them from two different points of view. First, we will discuss limits from an inexact intuitive point of view. Then, when you are familiar with the idea, we will give a precise mathematical definition of a limit.

Go to 98.

98

Here is a little bit of mathematical shorthand which will be useful in this section.

Suppose a variable x has values lying in an interval with the following properties:

1) The interval is centered at some number a.

2) The difference between x and a must be less than another number B.

3) x cannot have the particular value a. (We will shortly see why we want to exclude this point.)

The above three statements can be summarized in the following way:

$|x - a| > 0$ (This statement means x cannot have the value a.)

$|x - a| < B$ (The magnitude of the difference between x and a is less than B.)

These relations can be combined in the single statement:

$$0 < |x - a| < B.$$

(If you need to review the symbols used here, see frame 20.)

The values of x which satisfy $0 < |x - a| < B$ are indicated by the interval along the x-axis shown in the figure.

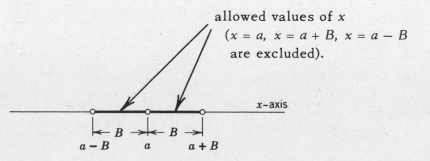

allowed values of x
($x = a$, $x = a + B$, $x = a - B$
are excluded).

Go to 99.

99

To begin our discussion of limits, we will look at an example. We are going to work with the equation $y = f(x) = x^2$ which is shown in the graph on the right. P represents the point on the curve corresponding to $x = 3$, $y = 9$.

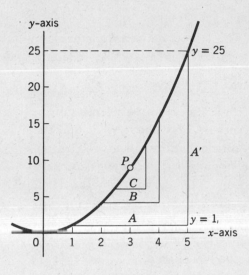

Let us concentrate on the behavior of y for values of x in an interval about $x = 3$. For reasons which we shall see shortly, it is important to exclude the particular point of interest, P, and to remind us of this, the point is encircled on the graph.

We start by considering values of y corresponding to values of x in an interval about $x = 3$ and lying between $x = 1$ and $x = 5$. With the notation of the last frame, this can be written as $0 < |x - 3| < 2$. The interval for x is shown by line A in the figure. The corresponding interval for y is shown by line A' and includes points between $y = 1$ and $y = 25$, except $y = 9$.

A smaller interval for x is shown by line B. Here $0 < |x - 3| < 1$, and the corresponding interval for y is $4 < y < 16$, with $y = 9$ excluded.

The interval for x shown by the line C is given by $0 < |x - 3| < 0.5$. Write the corresponding interval for y in the blank below, assuming $y = 9$ is excluded.

_____

In order to find the correct answer, go on to 100.

100

The interval for y which corresponds to $0 < |x - 3| < 0.5$ is

$$6.25 < y < 12.25$$

which you can easily check by substituting the values 2.5 and 3.5 for x in $y = x^2$ in order to find the values of y at either end point.

So far we have considered three successively smaller intervals of x about $x = 3$, and the corresponding intervals of y. Suppose we continue the process. The drawing on the right shows the plot $y = x^2$ for values of x between 2.9 and 3.1. (It is an enlarged piece of the graph in frame 99.) Three small intervals of x around $x = 3$ are shown along with the corresponding interval in y. The table below shows the values of y, corresponding to the boundaries of x at either end of the interval. (The last entry is for an interval too small to show on the drawing.)

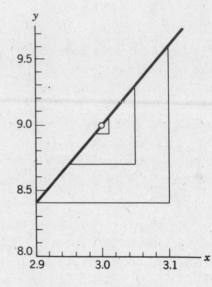

Interval of x	Corresponding interval of y
1–5	1–25
2–4	4–16
2.5–3.5	6.25–12.25
2.9–3.1	8.41–9.61
2.95–3.05	8.70–9.30
2.99–3.01	8.94–9.06
2.999–3.001	8.994–9.006

Go to 101.

101

We hope it is apparent from the discussion in the last two frames that as we diminish the interval for x around $x = 3$, the values for $y = x^2$ cluster more and more closely about $y = 9$. In fact, it appears that we can make the values for y cluster as closely as we please about $y = 9$ by merely limiting x to a sufficiently small interval about $x = 3$. Because this is true, we say that the *limit* of x^2, as x approaches 3, is 9, and we write this

$$\lim_{x \to 3} x^2 = 9.$$

Let's put this in more general terms.

If a function $f(x)$ is defined for values of x about some fixed number a, and if, as x is confined to smaller and smaller intervals about a, the values of $f(x)$ cluster more and more closely about some specific number L, the number L is called the *limit* of $f(x)$ as x approaches a. The statement that "the limit of $f(x)$ as x approaches a is L" is customarily abbreviated by

$$\lim_{x \to a} f(x) = L.$$

In the example at the top of the page $f(x) = x^2$, $a = 3$, and $L = 9$.

The important idea in the definition is that the intervals we use lie around the point of interest, a, but that the point itself is not included. In fact, $f(a)$, the value of the function at a, may be entirely different from $\lim_{x \to a} f(x)$, as we shall see.

Go to 102.

102

You may be wondering why we have been giving such a complicated discussion of an apparently simple problem. Why bother with $\lim_{x \to 3} x^2 = 9$ when it is obvious that $x^2 = 9$ for $x = 3$?

The reason is that often the value of a function for a particular $x = a$ is not defined, whereas the limit as x approaches a is perfectly well defined. For instance at $\theta = 0$ the function $\dfrac{\sin \theta}{\theta}$ has the value $\dfrac{0}{0}$, which is meaningless. When we get to frame 110 we shall see that

$$\lim_{\theta \to 0} \frac{\sin \theta}{\theta} = 1.$$

As another illustration consider

$$f(x) = \frac{x^2 - 1}{x - 1}.$$

For $x = 1$, $f(1) = \dfrac{1 - 1}{1 - 1} = \dfrac{0}{0}$, which is not defined. However we can divide by $x - 1$ *provided* x is not equal to 1, and we obtain

$$f(x) = \frac{x^2 - 1}{x - 1} = \frac{(x + 1)(x - 1)}{x - 1} = x + 1.$$

Therefore, even though $f(1)$ is not defined,

$$\lim_{x \to 1} f(x) = \lim_{x \to 1} (x + 1) = 2.$$

Formal justification of the last two steps is given in Appendix A2, along with a number of rules for handling limits. Do not read the Appendix now unless you are really interested.

We could also have obtained the above result graphically by studying the graph of the function in the neighborhood of $x = 1$ as we did in Frame 99.

Go to 103.

103

To see whether you have caught on, find the limit of the following slightly more complicated functions by procedures similar to the above: (You will probably have to work these out on paper. Both of them involve a little algebraic manipulation.)

(a) $\lim\limits_{x \to 0} \dfrac{(1 + x)^2 - 1}{x} = \boxed{1 \mid x \mid - 1 \mid 2}$

(b) $\lim\limits_{x \to 0} \dfrac{1 - (1 + x)^3}{x} = \boxed{1 \mid x \mid 3 \mid - 3}$

If right, go to 105.
Otherwise, go to 104.

104

Here are the solutions to the problems in 103:

(a) $\lim\limits_{x \to 0} \dfrac{(1 + x)^2 - 1}{x} = \lim\limits_{x \to 0} \dfrac{(1 + 2x + x^2) - 1}{x}$

$= \lim\limits_{x \to 0} \dfrac{2x + x^2}{x} = \lim\limits_{x \to 0} (2 + x) = \lim\limits_{x \to 0} 2 + \lim\limits_{x \to 0} x = 2$

(b) $\lim\limits_{x \to 0} \dfrac{1 - (1 + x)^3}{x} = \lim\limits_{x \to 0} \dfrac{1 - (1 + x)(1 + x)(1 + x)}{x}$

$= \lim\limits_{x \to 0} \dfrac{1 - (1 + 3x + 3x^2 + x^3)}{x} = \lim\limits_{x \to 0} (-3 - 3x - x^2)$

$= \lim\limits_{x \to 0} (-3) + \lim\limits_{x \to 0} (-3x) + \lim\limits_{x \to 0} (-x^2) = -3$

Again, if you would like justification of the steps used in these proofs, see Appendix A2.

Go to 105.

105

So far we have discussed limits only in an informal and intuitive way using expressions such as "confined to a smaller
and smaller interval" and "clustering more and more closely."
These expressions convey the intuitive meaning of a limit, but
they are not precise mathematical statements. Now we are ready
for a precise definition of a limit. [Since it is an almost universal
custom, in the definition of a limit we will use the Greek letters
δ (delta) and ϵ (epsilon).] Here we go.

Definition of a Limit.

Let $f(x)$ be defined for all x in an interval about $x = a$, but
not necessarily at $x = a$. If there is a number L such that to
each positive number ϵ there corresponds a positive number δ
such that

$$|f(x) - L| < \epsilon \qquad provided \qquad 0 < |x - a| < \delta$$

we say that L is the limit of $f(x)$ as x approaches a, and write

$$\lim_{x \to a} f(x) = L.$$

Go to 106.

Answers: (103) 2, -3.

106

The formal definition of a limit in frame 105 provides a clear basis for settling a dispute as to whether the limit exists and is L. Suppose we assert that $\lim\limits_{x \to a} f(x) = L$, and an opponent disagrees. As a first step, we tell him to pick a positive number ϵ, as small as he pleases, say 0.001, or if he wants to be difficult, 1000^{-1000}. Our task is to find some other number, δ, such that for all x in the interval $0 < |x - a| < \delta$, the difference between $f(x)$ and L is smaller than ϵ. If we can always do this, we win the argument — the limit exists and is L. These steps are illustrated for a particular function in the drawings below.

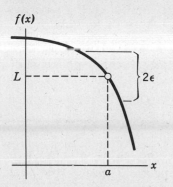   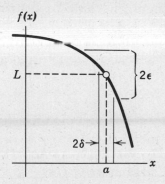

Our opponent has challenged us to find a δ to fit this ϵ.

Here is one choice of δ. Obviously, for all values of x in the interval shown, $f(x)$ will satisfy $|f(x) - L| < \epsilon$.

It may be that our opponent can find an ϵ such that we can never find a δ, no matter how small, that satisfies our requirement. In this case, he wins and $f(x)$ does not have the limit L. (In frame 114 we will come to an example of a function which does not have a limit.)

Our formal definition of a limit is clearly more precise than the expression in frame 101 that ''if x is confined to smaller and smaller intervals about a, the values of $f(x)$ cluster more and more closely about L.''

Go to 107.

107

In the examples we have studied so far, the function has been expressed by a single equation. However, this is not necessarily the case. Here is an example to show this.

$f(x) = 1$ for $x \neq 2$

$f(x) = 3$ for $x = 2$

(The symbol $\neq$ means not equal.)

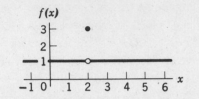

A suggestive sketch of this peculiar function is shown. You should be able to convince yourself that $\lim\limits_{x \to 2} f(x) = 1$ whereas $f(2) = 3$.

If you would like further explanation of this, go to 108.

Otherwise, go to 109.

108

For every value of x except $x = 2$, the value of $f(x) = 1$. Consequently, $f(x) - 1 = 0$ for all x except $x = 2$. Since 0 is less than the smallest positive number ϵ that your opponent could select, it follows from the definition of a limit that $\lim\limits_{x \to 2} f(x) = 1$, even though $f(2) = 3$.

Go to 109.

109

Here is another function which has a well-defined limit, but which can't be evaluated at the limiting point: Consider $f(x) = (1 + x)^{1/x}$. The value of $f(x)$ at $x = 0$ is quite puzzling. However, it is possible to find $\lim\limits_{x \to 0} (1 + x)^{1/x}$.

We will see later how to find this limit numerically. Its value turns out to be 2.718..... This quantity will be important in our study of logarithms and it is given a special symbol, e. Like π, e is irrational: that is, it is an unending and unrepeating decimal.

The procedure for evaluating e is discussed in Appendix A8, which you can read if you are curious.

Go to 110.

110

The actual procedure for finding a limit varies from problem to problem. There are a number of theorems for finding the limits of simple functions in Appendix A2, which you should read if you are interested. The result mentioned earlier,

$$\lim_{\theta \to 0} \frac{\sin \theta}{\theta} = 1$$

is proven in Appendix A3. (The limit is 1 only if θ is measured in radians.)

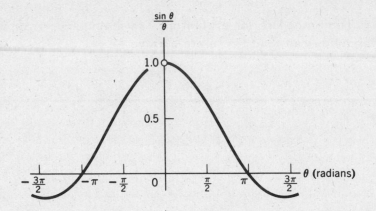

You can see that this result is reasonable by graphing the function $\frac{\sin \theta}{\theta}$ as shown above. We can easily do this with the aid of trigonometric tables, except, of course for $\theta = 0$. From the figure it should be reasonable that

$$\lim_{\theta \to 0} \frac{\sin \theta}{\theta} = 1.$$

Go to 111.

111

So far in our discussion of limits we have been careful to neglect the actual value of $f(x)$ at the point of interest, a. In fact, $f(a)$ does not even need to be *defined* for the limit to exist. However, frequently $f(a)$ is defined. If this is so, and if in addition

$$\lim_{x \to a} f(x) = f(a)$$

then the function is said to be *continuous* at a. To summarize, fill in the blanks:

A function f is continuous at $x = a$ if

(1) $f(a)$ is _____
.

(2) $\lim_{x \to a} f(x) = $ _____

Check your answers in frame 112.

112

Here are the correct answers: A function f is continuous at $x = a$ if

(1) $f(a)$ is *defined*.

(2) $\lim\limits_{x \to a} f(x) = f(a).$

You can always draw the graph of a continuous function without lifting your pencil from the paper in the region of interest.   Try to determine whether each of the following functions is continuous at the point indicated.

(1) $f(x) = \dfrac{x^2 + 3}{9 - x^2}.$

   At $x = 3$, $f(x)$ is $\lceil$continuous | discontinuous$\rceil$.

(2) $f(x) = \begin{cases} 1, & x \geq 0 \\ 0, & x < 0 \end{cases}$

   At $x = 1$, $f(x)$ is $\lceil$continuous | discontinuous$\rceil$.

(3) $f(x) = |\, x \,|.$

   At $x = 0$, $f(x)$ is $\lceil$continuous | discontinuous$\rceil$.

(4) The function $f(x)$ described in frame 107.

   At $x = 2$, $f(x)$ is $\lceil$continuous | discontinuous$\rceil$.

*If you had correct answers to
all these, go to 114.*

*If you made any mistakes, or
want more explanation, go to 113.*

113

Here are the explanations of the problems in frame 112.

(1) At $x = 3$, $f(x) = \dfrac{x^2 + 3}{9 - x^2} = \dfrac{12}{0}$. This is an undefined expression and, therefore, the function is not continuous at $x = 3$.

(2) Here is a plot of the function given.

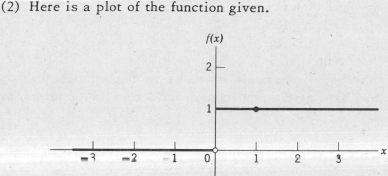

This function satisfies both conditions for continuity at $x = 1$, and is thus continuous there. (It is, however, discontinuous at $x = 0$.)

(3) Here is a plot of $f(x) = |x|$.

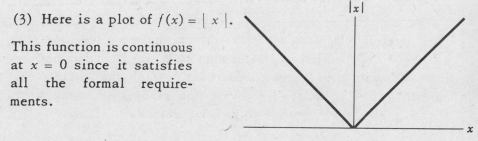

This function is continuous at $x = 0$ since it satisfies all the formal requirements.

(4) For the function described in frame 107, $f(x)$ has the value 3 at $x = 2$, while $\lim\limits_{x \to 2} f(x) = 1$. Since the value of the function and its limit are different at that point, it is discontinuous at that point.

Go to 114.

Answers: (112) discontinuous, continuous, continuous, discontinuous

114

Before leaving the subject of limits, perhaps we ought to look at some examples of functions which somewhere have no limit. One such function is that described in problem (2) of frame 113. The graph of the function is

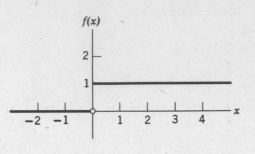

shown in the figure. We can show that this function has no limit at $x = 0$ by following the procedure described in the definition of a limit.

For purposes of illustration, suppose we guess that $\lim_{x \to 0} f(x) = 1$. Next, our opponent chooses a value for ϵ, say $1/4$. Now, for $|x - 0| < \delta$, where δ is *any* positive number,

$$|f(x) - 1| = \begin{cases} |1 - 1| = 0 & \text{if } x > 0 \\ |0 - 1| = 1 & \text{if } x < 0 \end{cases}$$

Therefore, for all negative values of x in the interval, $|f(x) - 1| = 1$, which is greater than $\epsilon = 1/4$. Thus 1 is not the limit. You should be able to convince yourself that there is *no* number L which satisfies the criterion since $f(x)$ changes by 1 when x goes from negative to positive values.

Go to 115.

115

Here is another example of a function which has no limit at a point. From the graph it is obvious that cot θ has no limit as $\theta \to 0$. Instead of clustering more and more closely to any number, L, the value of the function gets increasingly larger as $\theta \to 0$ in the direction shown by A, and increasingly more negative as $\theta \to 0$ in the direction shown by B.

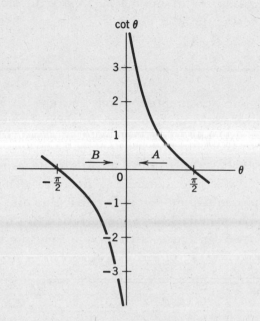

This concludes our study of the limit of a function for the present. If you would like some more practice with limits, see review problems 21 through 28 on page 276.

Now we are ready to go on to the next section.

Go to 116.

Section 2.   VELOCITY

116

We have been getting a little abstract, so before we get into differential calculus let's talk about something down to earth: motion, for instance. As a matter of fact, Leibniz and Newton invented calculus just because they were concerned with problems of motion, so it isn't a bad place to start. Besides, you already know quite a bit about it.

Go to 117.

117

Here is a warm-up problem which you should be able to do. In this problem, as in all problems in this chapter, the motion will be along a straight line.

A train travels away from us at a velocity v mph (miles per hour). At $t = 0$, it is at a distance S_0 from us. (The subscript on S_0 is to avoid confusion. S_0 is a particular distance and is a constant; S is a variable.) Write the equation for the distance the train is from us, S, in terms of time, t. (Take the unit of t to be hours.)

$S = $ _____

Go to 118 for the answer to this.

118

If you wrote $S = S_0 + vt$, you are correct. Go on to frame 119.

If your answer was not equivalent to the above, you should convince yourself that this answer is correct. Note that it yields $S = S_0$ when $t = 0$, as required. The equation is that of a straight line, and it might be a good idea to review the section on linear functions, Section 3 of Chapter I, p. 15 before continuing. When you are satisfied with this result, go to 119.

119

Here is a plot of the positions at different times of a train going in a straight line. Obviously, this represents a linear equation. Write the equation for the position of this train (in miles) in terms of time (in hours).

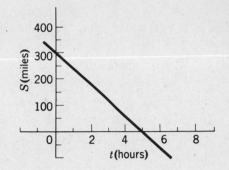

$S = $ _____ .

Find the velocity of the train from your equation.

$v = $ _____

Go to 120 for the correct answers.

120

Here are the answers to the questions in frame 120.

$$S = 60t + 300$$

$$v = -60 \text{ mph.}$$

The velocity is negative because S decreases with increasing time. If you would like some further discussion you should review frames 33 and 34.

Go to 121.

121

Here is another plot of position of a train travelling in a straight line.

The property of the line which represents the *velocity* of the train is the _____ of the line.

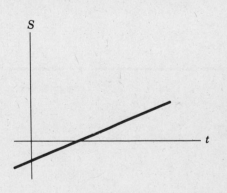

Go to 122 for the answer.

122

The property of the line which represents the velocity of the train is the *slope* of the line.

If you wrote this, go right on to 123.

If you wrote anything else, or nothing at all, then you may have forgotten what we reviewed back in section 3 of Chapter I. You should go over that section once again (particularly frames 33 and 34) and think about this problem before going on. At least convince yourself that the slope really represents the velocity.

Go to 123.

123

Here are plots of the positions vs. time of 6 different objects moving along straight lines. Which plot corresponds to the object that

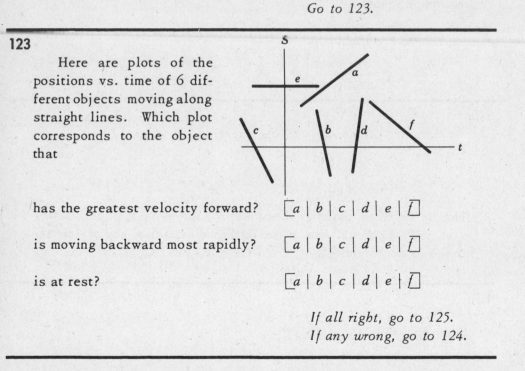

has the greatest velocity forward?   [*a* | *b* | *c* | *d* | *e* | *f*]

is moving backward most rapidly?   [*a* | *b* | *c* | *d* | *e* | *f*]

is at rest?   [*a* | *b* | *c* | *d* | *e* | *f*]

If all right, go to 125.
If any wrong, go to 124.

124

The velocity of the object is given by the slope of the plot of its distance against time.   Don't confuse the slope of a line with its location.

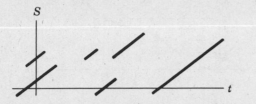 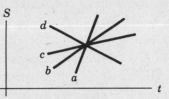

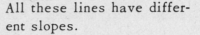

All the above lines have the same slope.

All these lines have different slopes.

A positive slope means that distance is increasing with time, which corresponds to a positive velocity.   Likewise, a negative slope means that distance is decreasing in time, which means the velocity is negative. If you need to review the idea of slope, look at frames 25–27 before continuing.

You should be able to answer these questions:

Which line in the figure above on the right has

negative slope? $\left[\, a \mid b \mid c \mid \bar{d}\, \right]$

greatest positive slope? $\left[\, a \mid b \mid c \mid \bar{d}\, \right]$

Go to 125.

Answers: (123) *d, b, e*

125

So far, the velocities we have considered have all been constant in time. But what if the velocity changes?

Here is a plot of the position of a car which is travelling with varying velocity along a straight line. In order to describe this motion, we introduce the *average velocity* $\bar{v}$

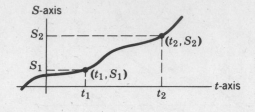

(read as "v bar"), which is the ratio of the net distance travelled to the time taken. For example, between the times t_1 and t_2 the car went a net distance $S_2 - S_1$, so $(S_2 - S_1)/(t_2 - t_1)$

was its _____   _____ during the time.

Go to 126.

126

The answer to frame 125 is

$(S_2 - S_1)/(t_2 - t_1)$ was its *average velocity* during the time.

(The single word "velocity" is not a correct answer.)

Go to 127.

127

In addition to defining the average velocity, $\bar{v}$, algebraically,

$$\bar{v} = \frac{S_2 - S_1}{t_2 - t_1},$$

we can interpret $\bar{v}$ graphically. If we draw a straight line between the points (t_1, S_1) and (t_2, S_2), then the average velocity is simply the *slope* of that line.

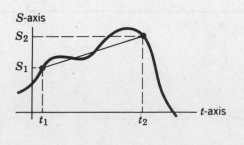

Go to 128.

Answers: (124) *d*, *a*

128

During which interval was the average velocity

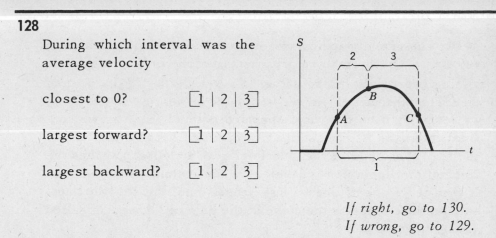

closest to 0?   ⊏1 | 2 | 3⊐

largest forward?   ⊏1 | 2 | 3⊐

largest backward?   ⊏1 | 2 | 3⊐

If right, go to 130.
If wrong, go to 129.

129

Since you missed the last problem, we'll analyze it in detail.

Here are straight lines drawn through the points *A, B, C*. Line (I) has a very small slope and corresponds to almost 0 velocity. Line (II) has positive slope, and line (III) has negative slope, corresponding to positive and negative average velocities, respectively.

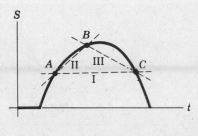

Go to 130.

130

We now extend our idea of velocity in a very important manner: instead of asking "what is the average velocity between time t_1 and t_2?", let us ask "what is the velocity at time t_1?" The velocity at a particular time is called the INSTANTANEOUS velocity. This is a new term, and we will give it a very precise definition shortly even though it may already be somewhat familiar to you.

Go to 131.

131

    We can give a graphical meaning to the idea of instantaneous velocity. The average velocity is the slope of a straight line joining two points on the curve, (t_1, S_1) and (t_2, S_2). To find the instantaneous velocity we want t_2 to be very close to t_1. As we let point B on the curve approach point A (i.e., as we consider intervals of time, starting at t_1 which become shorter and shorter), the slope of the line joining A and B approaches the slope of the line which is labeled l. The instantaneous velocity is then the *slope* of line l. In a sense, then, the straight line l has the same slope as the curve at the point A. Line l is called a *tangent* to the curve.

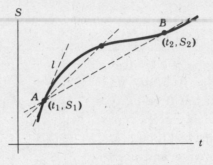

Go to 132.

132

    Here is where the idea of a limit becomes very important. If we draw a straight line through the given point A on the curve and some other point on the curve, B, and then let B get closer and closer to A, the slope of the straight line approaches a unique value, and can be identified with the *slope* of the curve at A. What we must do is consider the *limit* of the slope of the line through A and B as $B \longrightarrow A$.

Now, go to 133.

Answers: (128) 1, 2, 3

133

   We will now give a precise meaning to the intuitive idea of instantaneous velocity as the slope of a curve at a point. We start by considering the average velocity:

   $\bar{v} = (S_2 - S_1)/(t_2 - t_1) =$ the slope of the line connecting points 1 and 2.

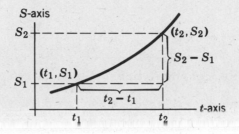

   As $t_2 \longrightarrow t_1$, the average velocity approaches the instantaneous velocity, i.e., $\bar{v} \longrightarrow v$ as $t_2 \longrightarrow t_1$, or

$$v = \lim_{t_2 \to t_1} \frac{S_2 - S_1}{t_2 - t_1}$$

Go to 134.

134

   Since the ideas presented in the last few frames are very important, let's summarize them.

   If a point moves from S_1 to S_2 during the time t_1 to t_2, then $(S_2 - S_1)/(t_2 - t_1)$ is the _____ _____ , $\bar{v}$.

   If we consider the limit of the average velocity as the averaging time goes to zero, the result is called the

_____ _____ , v.

   Now let's try to present these ideas in a neater form. If you can, write a formal definition of v in the blank space.

   $v =$

_____

Go to frame 135 for the answers.

135

The correct answers to frame 134 are the following:

If a point moves from S_1 to S_2 during the time t_1 to t_2, then $(S_2 - S \ /(t_2 - t_1)$ is the *average velocity*, $\bar{v}$.

If we consider the limit of the average velocity as the averaging time goes to zero, the result is called the *instantaneous velocity, v.*

$$v = \lim_{t_2 \to t_1} \frac{S_2 - S_1}{t_2 - t_1}$$

If you wrote this, congratulations! Go on to 136.

If you wrote something different, go back to frame 133 and work your way to this frame once more.

Then go on to 136.

136

To make the notation more succinct, we let $S_2 = S_1 + \Delta S$, $t_2 = t_1 + \Delta t$.   That is, the point moves distance ΔS in time Δt. (ΔS is a single symbol read as "delta S"; it does not mean $\Delta \times S$.) Although this notation may be new, it is worth the effort to get used to it since it saves lots of writing. Since $S_2 = S_1 + \Delta S$, then $\Delta S = S_2 - S_1$ by definition. Similarly, $\Delta t = t_2 - t_1$. In general, $\Delta x = x_2 - x_1$ where x is any variable, and x_2 and x_1 are any two given values of x.

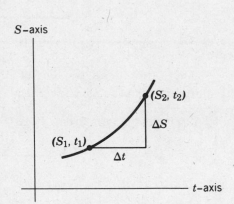

Similarly, if $y = f(x)$, $\Delta y = y_2 - y_1 = f(x_2) - f(x_1) = f(x_1 + \Delta x) - f(x_1)$.

With this notation, our definition of instantaneous velocity is

$v = $

_____

*Go to 137 to find
the correct answer.*

137

If you wrote

$$v = \lim_{\Delta t \to 0} \frac{\Delta S}{\Delta t},$$

you are really catching on. Go ahead to frame 138.

If you missed this, you need a little review. Study frames 134–136 and then go to 138.

138

Now we are going to apply the idea of instantaneous velocity by analyzing an example step by step. Later on we will find short cuts for doing this.

Suppose that we are given the following expression relating position and time.

$$S = f(t) = kt^2 \ (k \text{ is a constant})$$

Here are the steps for evaluating v:

$$f(t + \Delta t) = k[t + \Delta t]^2 = k[t^2 + 2t\Delta t + (\Delta t)^2]$$

$$\Delta S = f(t + \Delta t) - f(t) = k[t^2 + 2t\Delta t + (\Delta t)^2] - kt^2$$
$$= k[2t\Delta t + (\Delta t)^2]$$

$$\frac{\Delta S}{\Delta t} = \frac{k[2t\Delta t + (\Delta t)^2]}{\Delta t} = 2kt + k\Delta t$$

$$v = \lim_{\Delta t \to 0} \frac{\Delta S}{\Delta t} = \lim_{\Delta t \to 0} [2kt + k\Delta t] = 2kt.$$

A simpler problem for you to try is in the next frame.

Go to 139.

139

Suppose we are given that $S = f(t) = v_0 t + S_0$.

The problem is to find the instantaneous velocity from our definition.

In time Δt the point moves distance ΔS.

$\Delta S = $ _____

$v = \lim\limits_{\Delta t \to 0} \dfrac{\Delta S}{\Delta t} = $ _____

Write in the answers
and go to 140.

140

If you wrote

$\Delta S = v_0 \Delta t$

and

$v = \lim\limits_{\Delta t \to 0} \dfrac{\Delta S}{\Delta t} = v_0,$

you are correct and can skip on to frame 142.

If you wrote something different, study the detailed explanation in frame 141.

141

Here is the correct procedure. Since $S = f(t) = v_0 t + S_0$,

$\Delta S = f(t + \Delta t) - f(t)$

$\quad = v_0 [t + \Delta t] + S_0 - [v_0 t + S_0]$

$\quad = v_0 \Delta t$

$\lim\limits_{\Delta t \to 0} \dfrac{\Delta S}{\Delta t} = \lim\limits_{\Delta t \to 0} \dfrac{v_0 \Delta t}{\Delta t} = \lim\limits_{\Delta t \to 0} v_0 = v_0$

The instantaneous velocity and the average velocity are the same in this case, since the velocity is a constant, v_0.

Go to frame 142.

142

Here is a problem for you to work out. Suppose the position of an object is given by

$$S = f(t) = kt^2 + lt + S_0,$$

where k, l and S_0 are constants. Find v.

$$v = \lim_{\Delta t \to 0} \frac{\Delta S}{\Delta t} = \underline{\hspace{3cm}}.$$

To check your answer,
go to 143.

143

The answer is

$$v = 2kt + l.$$

If you obtained this result, go on to the next section which starts with frame 146.

Otherwise,

go to 144.

144

Here is how to work the problem in frame 142.

$$f(t) = kt^2 + lt + S_0$$

$$f(t + \Delta t) = k[t + \Delta t]^2 + l[t + \Delta t] + S_0$$

$$= k[t^2 + 2t\Delta t + (\Delta t)^2] + l[t + \Delta t] + S_0$$

$$\Delta S = f(t + \Delta t) - f(t) = k[2t\Delta t + (\Delta t)^2] + l\Delta t$$

$$v = \lim_{\Delta t \to 0} \frac{\Delta S}{\Delta t} = \lim_{\Delta t \to 0} \left\{ \frac{k[2t\Delta t + (\Delta t)^2] + l\Delta t}{\Delta t} \right\}$$

$$= \lim_{\Delta t \to 0} \{k[2t + \Delta t] + l\} = 2kt + l$$

Now try this problem:

If $S = At^3$, where A is a constant, find v.

Answer: _____

*To check your solution,
go to 145.*

145

Here is the answer: $v = 3At^2$. Go right on to frame 146 unless you would like to see the solution, in which case you should continue here.

$$S = At^3$$

$$\Delta S = A[t + \Delta t]^3 - At^3$$

$$= A[t^3 + 3t^2\Delta t + 3t(\Delta t)^2 + (\Delta t)^3] - At^3$$

$$= 3At^2\Delta t + 3At(\Delta t)^2 + A(\Delta t)^3$$

$$v = \lim_{\Delta t \to 0} \frac{\Delta S}{\Delta t} = \lim_{\Delta t \to 0} [3At^2 + 3At\Delta t + A(\Delta t)^2] = 3At^2.$$

*Go to the next section,
frame 146.*

Section 3. DERIVATIVES

146

In this section we will generalize our results on velocity. This will lead us to the idea of a *derivative*, which is at the very heart of differential calculus.

Go to 147.

147

Fill in the blanks below.

When we write $S = f(t)$, we are stating that position depends on time. Here position is the dependent variable and time is the

_____ variable.

The velocity is the rate of change of position with respect to time. By this we mean that velocity is (give the formal definition again):

$v =$

_____

Go to frame 148 for the correct answers.

148

In the last frame you should have written:

.... time is the *independent* variable,

and

$$v = \lim_{\Delta t \to 0} \frac{\Delta S}{\Delta t}$$

In any case, go on to 149.

149

Let us consider any continuous function defined by, say, $y = f(x)$. Now y is our dependent variable, and x is our independent variable. If we ask "at what rate does y change as x changes?", we can find the answer by taking the following limit:

$$\text{rate of change of } y \text{ with respect to } x = \lim_{\Delta x \to 0} \frac{\Delta y}{\Delta x}.$$

Go to 150.

150

You can give a geometrical meaning to $\lim\limits_{\Delta x \to 0} \dfrac{\Delta y}{\Delta x}$, where $y = f(x)$. To do so, fill in the blanks.

Geometrically, $\lim\limits_{\Delta x \to 0} \dfrac{\Delta y}{\Delta x}$ can be found by drawing a straight line through the point (x, y) and the point (_____ , _____) as shown. The slope of that line is given by $\dfrac{\Delta y}{\Delta x}$, and $\lim\limits_{\Delta x \to 0} \dfrac{\Delta y}{\Delta x}$ is the

_____ of the curve at (x, y).

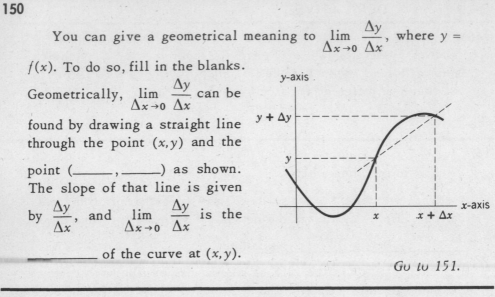

Go to 151.

151

The correct insertions for frame 150 are:

$(x + \Delta x,\ y + \Delta y)$,

$\lim\limits_{\Delta x \to 0} \dfrac{\Delta y}{\Delta x}$ is the *slope* of the curve at (x, y).

(If you would like to see a discussion of this, review frame 131 before continuing.)

Go to 152.

152

Another way of writing $\dfrac{\Delta y}{\Delta x}$ is

$\dfrac{y_2 - y_1}{x_2 - x_1}$, or $\dfrac{f(x_2) - f(x_1)}{x_2 - x_1}$.

If the notation used here still seems unfamiliar, you should review frame 136 before proceeding.

Go to 153.

153

Let's review just once more.

If we want to know how y changes as x changes, we find out by taking the following limit:

_____

*Fill in the blank and go
on to 154.*

154

The correct answer to frame 153 is

$$\lim_{\Delta x \to 0} \frac{\Delta y}{\Delta x}, \text{ or } \lim_{x_2 \to x_1} \frac{y_2 - y_1}{x_2 - x_1}.$$

If you were correct, good! Go on to 155.

If you missed this, *go back to 149.*

155

Because the quantity $\lim_{\Delta x \to 0} \frac{\Delta y}{\Delta x}$ is so useful, we give it a special name and a special symbol.

$\lim_{\Delta x \to 0} \frac{\Delta y}{\Delta x}$ is called the *derivative* of y with respect to x, and

it is written with the special symbol $\frac{dy}{dx}$.

$$\boxed{\frac{dy}{dx} = \lim_{\Delta x \to 0} \frac{\Delta y}{\Delta x}}$$

Once again: $\frac{dy}{dx}$ is the _____ of _____

with respect to _____ .

*Go to 156 for the
correct answer.*

156

The correct statement is:

$\dfrac{dy}{dx}$ is the *derivative* of y with respect to x.

Even though $\dfrac{dy}{dx}$, looks like a fraction, it is defined here as a complete symbol which stands for $\lim\limits_{\Delta x \to 0} \dfrac{\Delta y}{\Delta x}$. The symbol is often read as "dee y dee x" or "dee y by dee x." Sometimes $\dfrac{dy}{dx}$ is written as y', but we will always use $\dfrac{dy}{dx}$.

We can apply this definition to the notion of velocity discussed previously. Since velocity is the rate of change of position with respect to time, velocity is the *derivative* of position with respect to time.

Go to 157.

157

Let's state the definition of a derivative using different variables. Suppose z is some independent variable, and q depends on z. Then the derivative of q with respect to z is

$$\dfrac{dq}{dz} = \underline{\hspace{3cm}}.$$

(Give formal definition.)

For the right answer,
go to 158.

158

Your answer should have been

$$\dfrac{dq}{dz} = \lim\limits_{\Delta z \to 0} \dfrac{\Delta q}{\Delta z}.$$

If so, go to 159.

If not, go back to frame 155
and try again.

159

For convenience of notation, the symbol $\dfrac{dy}{dx}$ is sometimes

written $\dfrac{d}{dx}(y)$. Here is an example of alternative ways to write

$\dfrac{dy}{dx}$. If $y = x^3 + 3$,

$$\frac{dy}{dx} = \frac{d(x^3 + 3)}{dx} = \frac{d}{dx}(x^3 + 3).$$

Similarly,

$$\frac{d(\theta^2 \sin \theta)}{d\theta} = \frac{d}{d\theta}(\theta^2 \sin \theta).$$

(θ is simply another variable here.   Angle is just as good a variable as, say, distance.)

*On to Section 4,
frame 160.*

160

We have just learned the formal definition of a derivative. Graphically, the derivative of a function $f(x)$ at some value of x is equivalent to the slope of a straight line which is tangent to the graph of the function at that point. Our chief concern in the rest of this chapter will be to find methods for evaluating derivatives of different functions.   However, in doing this it is very helpful to have some sort of intuitive idea of how the derivative behaves, and we can obtain this by looking at the graph of the function. If the graph has a steep positive slope, the derivative is large and positive. If the graph has a slight slope downwards, the derivative is small and negative. In this section we will get some practice putting to use such qualitative ideas as these, and in the following sections we will learn how to obtain derivatives precisely.

Go to 161.

161

Here is a plot of the simple function $y = x$. Be-

low it we have plotted $\dfrac{dy}{dx}$.

Since the slope of y is positive and constant,

$\dfrac{dy}{dx}$ is a positive constant.

(In fact, as we already,

know, $\dfrac{d}{dx}(x) = 1$.)

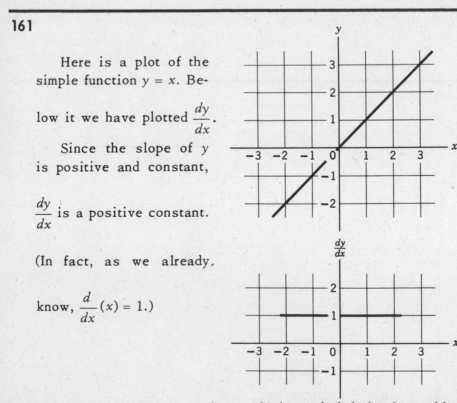

Now, go to frame 162 for a slightly harder problem.

162

Here is a plot of $y = |x|$. (If you have forgotten the definition of $|x|$, see frame 20.) On the coordinates below, sketch $\dfrac{dy}{dx}$.

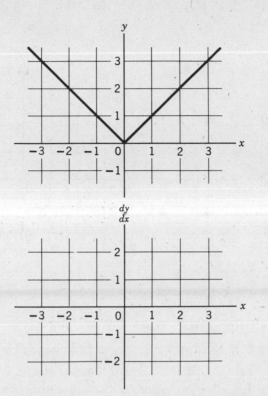

For the correct answer,
go to 163.

163

Here are sketches of $y = |x|$ and $\dfrac{dy}{dx}$. If you drew this correctly, go on to 164. If you made a mistake or want further explanation, continue here.

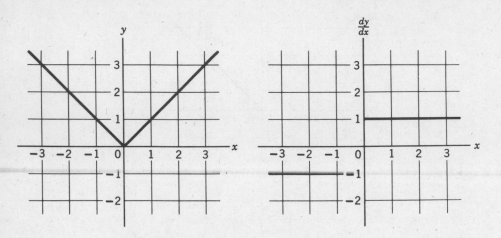

As you can see from the graph, $y = |x| = x$ for $x > 0$. So for $x > 0$ the problem is identical to that in frame 161, and $\dfrac{dy}{dx} = 1$.

However, for $x < 0$, the slope of $|x|$ is negative and is easily seen to be -1. At $x = 0$, $|x| = x = 0$ and the slope is undefined, for it has the value $+1$ if we approach 0 along the positive x-axis and has the value -1 if we approach 0 along the negative x-axis. Therefore, $\dfrac{d}{dx}|x|$ is discontinuous at $x = 0$. (The function x is continuous at this point, but the "break" in its slope at $x = 0$ causes a discontinuity in the derivative.)

Go to 164.

164

Here is the graph of a function $y = f(x)$. Sketch its derivative in the space provided below. (The sketch does not need to be exact — just show the general features of $\dfrac{dy}{dx}$.)

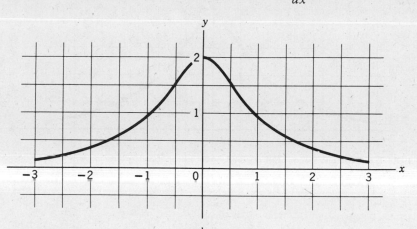

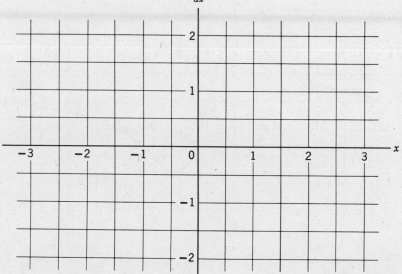

See frame 165 for the correct answer.

165

Here is the function and its derivative. If you drew a sketch of $\dfrac{dy}{dx}$ similar to that shown, go to 166. Otherwise, read on.

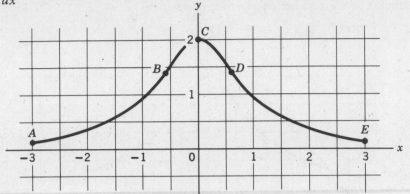

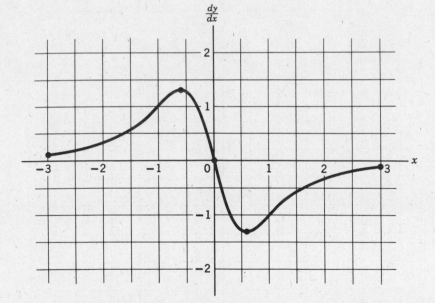

To see that the plot of dy/dx is reasonable, let us estimate dy/dx for several values of x. At the point C the graph has 0 slope, so dy/dx is 0. At B, y is increasing rapidly, so dy/dx is positive. At D, y is decreasing rapidly, so dy/dx is negative. At A and E the slope is small and dy/dx is close to 0. These values of dy/dx are enough to suggest its general behavior.

Go to 166.

166

Let's look at the behavior of $\dfrac{dy}{dx}$ graphically for one more function. Here the plot of y and x is a semicircle. In the space below, make a rough sketch of $\dfrac{dy}{dx}$ for the interval illustrated.

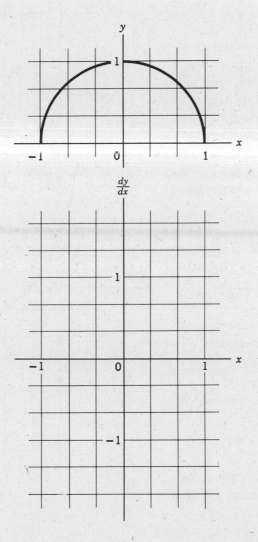

*Go to 167 for the
correct answer.*

167

Here are the plots of y and $\dfrac{dy}{dx}$. Read on if you would like further discussion of this. Otherwise, go to 168.

The slope of the semi-circle does not behave nicely at the extreme values of x, so let's start by looking at $x = 0$. If we draw a line tangent to the curve at $x = 0$, it will be parallel to the x-axis, so the curve has 0 slope. Thus, $\dfrac{dy}{dx} = 0$ at $x = 0$. For $x > 0$, a line tangent to the curve has negative slope, so $\dfrac{dy}{dx} < 0$. As x approaches 1 the tangent becomes i n c r e a s i n g l y steep, and $\dfrac{dy}{dx}$ becomes increasingly negative. In fact, as $x \to 1$, $\dfrac{dy}{dx} \to -\infty$.

From this discussion it should be easy to find $\dfrac{dy}{dx}$ for $x < 0$.

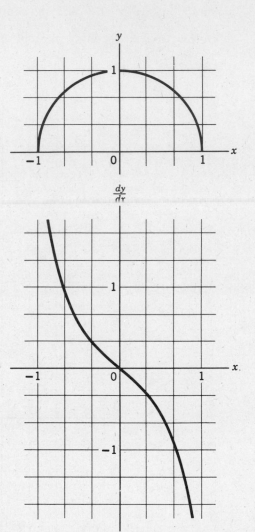

Go to 168.

168

If you understand all the examples in this section, skip on to the next section. However, if you would like a little more practice, try sketching the derivatives for each function shown. The correct sketches are given in frame 169 without any discussion.

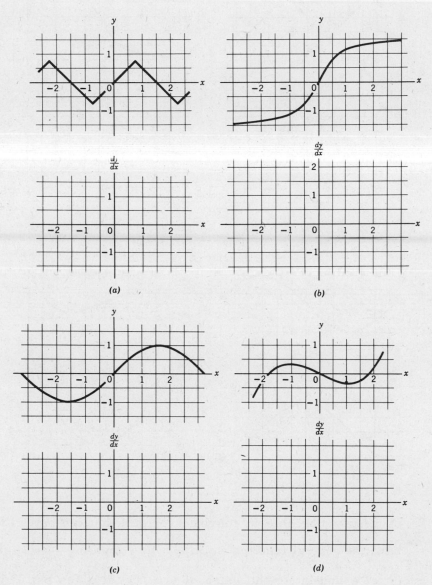

(a)   (b)

(c)   (d)

For the correct sketches, go to frame 169.

169

Here are the solutions to the problems in frame 168.

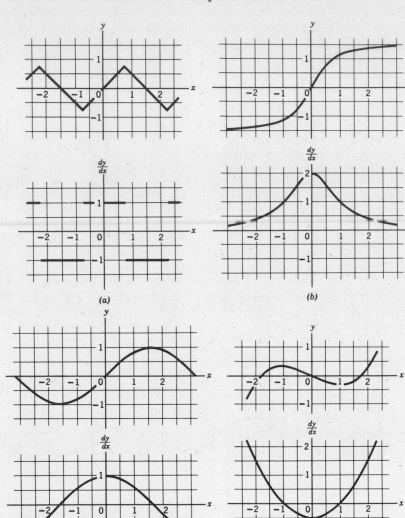

(a)   (b)

(c)   (d)

You should be able to convince yourself that the curves for $\dfrac{dy}{dx}$ have the general features we expect by comparing $\dfrac{dy}{dx}$ with the slope of a tangent to the graph of $y = f(x)$ at a few particular values of x.

Go to the next section, frame 170.

Section 5.   DIFFERENTIATION

170

We have accomplished a great deal so far in this chapter. In fact, all the really important new ideas involved in differential calculus have been introduced — limits, slopes of curves, and derivatives — and you are equipped in principle to solve a wide variety of problems. However, the process of applying the fundamental definition of derivative to each problem as it comes along would be very time consuming. It would also be a great waste of time since there are numerous rules and tricks for differentiating apparently complicated functions in a few short steps. You will learn the most important of these rules in the following sections. You will also learn how to differentiate a few functions which occur so often that it is useful to know their derivatives by heart. These include some of the trigonometric functions and logarithms and exponentials. The remaining sections cover some special topics, as well as applications of differential calculus to a few problems. By the end of this chapter you should be able to use differential calculus for many applications. Well, let's get going!

On to 171.

171

Can you find the derivative of the following simple function?

$y = a$ (a is a constant)

$\dfrac{dy}{dx} = \left[\, 1 \mid x \mid a \mid 0 \mid \text{none of these} \,\right]$

If right, go to 173.

If wrong, go to 172.

172

To find $\dfrac{dy}{dx}$, we go back to the definition $\dfrac{dy}{dx} = \lim\limits_{\Delta x \to 0} \dfrac{\Delta y}{\Delta x}$. If $y = a$,

$$\frac{\Delta y}{\Delta x} = \frac{f(x + \Delta x) - f(x)}{\Delta x} = \frac{a - a}{\Delta x} = 0.$$

(Remember that the meaning of $f(x + \Delta x)$ is f evaluated at $x + \Delta x$.)

$$\lim_{\Delta x \to 0} \frac{\Delta y}{\Delta x} = \lim_{\Delta x \to 0} 0 = 0.$$

Since $\dfrac{dy}{dx} = 0$, the plot of y in terms of x has 0 *slope*. (Example 4 in frame 32 shows this graphically.)

Go to 173.

173

You have just seen that the derivative of a constant is 0.

Now, try to find the derivative of this function:

$y = ax, \quad a = \text{constant}$

$\dfrac{dy}{dx} = \begin{bmatrix} 1 \mid x \mid a \mid 0 \mid ax \mid \text{none of these} \end{bmatrix}$

If right, skip to 175.

If wrong, go to 174.

174

Here is the correct procedure.

$f(x) = ax,$

$f(x + \Delta x) = a[x + \Delta x] = ax + a\Delta x$

so $\Delta y = f(x + \Delta x) - f(x) = [ax + a\Delta x] - ax = a\Delta x.$

Therefore

$$\lim_{\Delta x \to 0} \frac{\Delta y}{\Delta x} = \lim_{\Delta x \to 0} \frac{a\Delta x}{\Delta x} = a.$$

Now try to find the derivative of the function $y = x$.

$$\frac{dy}{dx} = \begin{bmatrix} 1 \mid 0 \mid a \mid -1 \mid x \end{bmatrix}$$

If correct, go to 175.

If wrong, note that this problem is just a special case of frame 173. Try again and then go to 175.

Answer: (171) 0

175

Now we are going to find the derivative of a quadratic function. Suppose

$$y = f(x) = x^2.$$

What is $\dfrac{dy}{dx}$?

You should be able to work this out from the definition of the derivative. Choose the correct answer;

$$\frac{dy}{dx} = \boxed{1 \mid x \mid 0 \mid x^2 \mid 2x}.$$

If right, go to 177,
Otherwise, go to 176.

176

Let us recall the definition of the derivative

$$\frac{dy}{dx} = \lim_{\Delta x \to 0} \frac{f(x + \Delta x) - f(x)}{\Delta x}.$$

In this case, $f(x + \Delta x) = [x + \Delta x]^2 = x^2 + 2x\Delta x + (\Delta x)^2$

so $\displaystyle \lim_{\Delta x \to 0} \frac{f(x + \Delta x) - f(x)}{\Delta x} = \lim_{\Delta x \to 0} \frac{[x^2 + 2x\Delta x + (\Delta x)^2] - x^2}{\Delta x}$

$$= \lim_{\Delta x \to 0} \frac{2x\Delta x + (\Delta x)^2}{\Delta x} = \lim_{\Delta x \to 0} (2x + \Delta x) = 2x,$$

so $\dfrac{dy}{dx} = 2x.$

Go to 177.

Answer: (173) *a*;   (174) 1

177

We have derived the result that $\dfrac{d}{dx}(x^2) = 2x$. To illustrate this, a graph of $y = x^2$ is drawn in the figure. Since the slope of the curve at a point is simply the derivative at that point, each of the straight lines tangent to the curve has a slope equal to the derivative evaluated at the point of tangency.

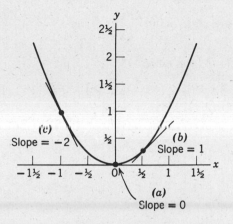

Line (a) is tangent through the origin, and has a slope of $2 \times (0) = 0$. Line (b) passes through the point $x = 1/2$, and has slope $2 \times (1/2) = 1$. Line (c) passes through the point $x = -1$, and has slope $2 \times (-1) = -2$.

Go to 178.

178

Here is a problem which summarizes the results we have had so far in this section (with a tiny bit of new material).

If $y = 3x^2 + 7x + 2$

Find $\dfrac{dy}{dx}$

Answer: $\dfrac{dy}{dx} =$ _____

See frame 179 for the correct answer.

Answer: (175) $2x$

179

If $y = 3x^2 + 7x + 2$, then $\dfrac{dy}{dx} = 6x + 7$.

Congratulations if you got this answer. Go on to 180. Otherwise, read below.

After you have finished this chapter you will know several shortcuts for evaluating this derivative. However, right now we will use the basic definition: $\dfrac{dy}{dx} = \lim\limits_{\Delta x \to 0} \dfrac{f(x + \Delta x) - f(x)}{\Delta x}$. Since

$f(x) = 3x^2 + 7x + 2$, we have

$f(x + \Delta x) = 3\left[x^2 + 2x\Delta x + (\Delta x)^2\right] + 7\left[x + \Delta x\right] + 2$

$f(x + \Delta x) - f(x) = 6x\Delta x + 3\Delta x^2 + 7\Delta x$

so $\dfrac{dy}{dx} = \lim\limits_{\Delta x \to 0}\left[\dfrac{6x\Delta x + 3\Delta x^2 + 7\Delta x}{\Delta x}\right] = \lim\limits_{\Delta x \to 0}\left[6x + 3\Delta x + 7\right]$

$= 6x + 7$.

Go to 180.

180

Now that we have found the derivatives of x and x^2, our next step is to find the derivative of x^n, where n is any number. We will state the rule here, but you can look in Appendix A4 if you would like to see how it is derived.

The result is

$$\boxed{\dfrac{dx^n}{dx} = nx^{n-1}}$$

This important result holds for all values of n: positive, negative, integral, fractional, irrational, etc. Note that our previous result, $\dfrac{d}{dx}(x^2) = 2x$, is the particular case of this when $n = 2$.

Go to 181.

181

Now for a few applications.

Find $\dfrac{dy}{dx}$ for each of the following functions.

$y = x^3$ $\dfrac{dy}{dx} = \boxed{3x^3 \mid 3x^2 \mid 2x^3 \mid x^2}$

$y = x^{-7}$ $\dfrac{dy}{dx} = \boxed{-7x^{-6} \mid 7x^{-7} \mid -7x^{-8} \mid -6x^{-7}}$

$y = \dfrac{1}{x^2}$ $\dfrac{dy}{dx} = \boxed{-2x \mid 2/x \mid -2/x^3}$

If all these were correct,
go to 183.

If you made any errors,
go to 182 to see how
to work the problems.

182

These problems should not have been difficult. They depend directly on the rule in frame 180. Here are the details.

We use our general rule: $\dfrac{d}{dx}\,x^n = nx^{n-1}$.

$y = x^3$;   in this case $n = 3$, so

$$\frac{d(x^3)}{dx} = 3x^{(3-1)} = 3x^2$$

$y = x^{-7}$; here $n = -7$, so

$$\frac{d(x^{-7})}{dx} = -7x^{(-7-1)} = -7x^{-8}$$

$y = 1/x^2 = x^{-2}$; here $n = -2$, so

$$\frac{d(1/x^2)}{dx} = -2x^{(-2-1)} = -2x^{-3} = -2/x^3$$

Now try these problems:

$$y = \frac{1}{x}, \qquad \frac{dy}{dx} = \left[1 + \frac{1}{x} \ \middle| \ -\frac{1}{x} \ \middle| \ -\frac{1}{x^2} \ \middle| \ 2\right]$$

$$y = \frac{-1}{3}\,x^{-3}, \ \frac{dy}{dx} = \left[x^{-4} \ \middle| \ -3x^{-4} \ \middle| \ \frac{-1}{4}\,x^{-2} \ \middle| \ +x^{-2}\right].$$

If right, go on to 183.

If wrong, go back to 180 and continue from there.

Answers: (181) $3x^2$, $-7x^{-8}$, $-2/x^3$

183

Here is another application.

If $y = x^{\frac{1}{2}}$ find $\dfrac{dy}{dx}$.

The answer is: $\left[x^{-\frac{1}{2}} \mid \dfrac{1}{2} x^{-\frac{1}{2}} \mid \dfrac{1}{2} x \mid \text{none of these} \right]$.

If right, go to Section 6,
frame 185.

If wrong, go to 184.

184

The rule $\dfrac{dx^n}{dx} = nx^{n-1}$ is true for any value of n.

In this case, $n = \dfrac{1}{2}$,

$$\frac{d}{dx} x^{\frac{1}{2}} = \frac{1}{2} x^{(\frac{1}{2} - 1)} = \frac{1}{2} x^{-\frac{1}{2}}.$$

Try this problem:

$$\frac{d}{dx} (x^{2/3}) = \left[x^{-1/3} \; \frac{2}{3} x^{-2/3} \mid \frac{2}{3} x^{-1/3} \mid x^{5/3} \right].$$

Go to Section 6, frame 185.

Answers: (182) $-1/x^2$, $\;x^{-4}$

Section 6.   SOME RULES FOR DIFFERENTIATION

185

In this section we are going to learn a number of shortcut rules for differentiation without having to go all the way back to the definition of the derivative each time. Some of these rules are derived here, while others are derived in Appendix A.

For the rest of this section, we will let $u(x)$ and $v(x)$ stand for any two variables that depend on x.

Go to 186.

186

Our first rule will let us evaluate the derivative of the sum of u and v, in terms of their derivatives. We will derive the rule here.

Let $y = u(x) + v(x)$

Then $\dfrac{dy}{dx} = \lim_{\Delta x \to 0} \; [u(x + \Delta x) + v(x + \Delta x) - u(x) - v(x)] \; \dfrac{1}{\Delta x}$

$= \lim_{\Delta x \to 0} \; [u(x + \Delta x) - u(x)] \; \dfrac{1}{\Delta x} +$

$\lim_{\Delta x \to 0} \; [v(x + \Delta x) - v(x)] \; \dfrac{1}{\Delta x}$

$= \dfrac{du}{dx} + \dfrac{dv}{dx}.$

Hence

$$\boxed{\dfrac{d}{dx}[u + v] = \dfrac{du}{dx} + \dfrac{dv}{dx}}$$

If you would like a rigorous justification of the manipulation of the limits in the above proof, see Appendix A2.

Go to 187.

Answers: (183) $\dfrac{1}{2} x^{-\frac{1}{2}}$; (184) $\dfrac{2}{3} x^{-1/3}$

187

Now let's put the above rule to use by computing the derivative of the following function (you will also have to use some results from section 5):

$$y = x^4 + 8x^3$$

$$\frac{dy}{dx} = \underline{\hspace{5cm}}$$

For the correct answer,
go to frame 188.

188

The correct answer to the question in frame 187 is

$$\frac{d}{dx}[x^4 + 8x^3] = 4x^3 + 24x^2.$$

If you got this answer, go to frame 189.

Otherwise, continue here to find your mistake.

Our problem is to find the derivative of the sum of two functions. To make use of the rule in frame 186 in the notation used there, suppose we let $u = x^4$, $v = 8x^3$.

Then $\frac{d}{dx}[u + v] = \frac{d}{dx}[x^4 + 8x^3] = \frac{d}{dx}[x^4] + \frac{d}{dx}[8x^3]$.

You should be able to evaluate these two derivatives from the result of the last section:

$$\frac{d}{dx}[x^4] = 4x^3, \qquad \frac{d}{dx}[8x^3] = 24x^2.$$

Hence, $\frac{d}{dx}[x^4 + 8x^3] = 4x^3 + 24x^2.$

Go to 189.

189

Now that we can differentiate the sum of two variables, our next task is to learn how to differentiate a product, for instance, $u(x) \, v(x)$. We want to express $\dfrac{d}{dx}[uv]$ in terms of $\dfrac{du}{dx}$ and $\dfrac{dv}{dx}$. The result will be stated here. Look in Appendix A6 if you want to see how it is derived. The rule, sometimes called the *product rule,* is

$$\frac{d}{dx}[uv] = u\,\frac{dv}{dx} + v\,\frac{du}{dx}$$

Go to 190.

190

Here is an example in which the *product rule* is used. Suppose $y = [x^5 + 7]\,[x^3 + 17x]$. The problem is to find $\dfrac{dy}{dx}$. If we let $u = x^5 + 7$, $v = x^3 + 17x$, then $y = uv$.

$$\frac{dy}{dx} = \frac{d}{dx}[uv] = u\,\frac{dv}{dx} + v\,\frac{du}{dx}.$$

Since $\dfrac{du}{dx} = 5x^4$ and $\dfrac{dv}{dx} = 3x^2 + 17$, our result is

$$\frac{dy}{dx} = [x^5 + 7]\,[3x^2 + 17] + [x^3 + 17x]\,[5x^4].$$

By using the product rule, we can derive in another way a result we have already found: $\dfrac{d}{dx}[x^2] = 2x$. If we let $u = x$, and $v = x$, then the product rule tells us that

$$\frac{d}{dx}\,x^2 = x\,\frac{dx}{dx} + x\,\frac{dx}{dx} = 2x.$$

Go to 191.

191

Now you try to evaluate a derivative using the product rule. Find

$$\frac{d}{dx}[(3x + 7)(4x^2 + 6x)].$$

Answer: _____

Go to 192 for the solution.

192

The answer to the problem is:

$$(3x + 7)(8x + 6) + (4x^2 + 6x)(3).$$

If you obtained this or an equivalent result, go on to 194. Otherwise, read below.

The problem is to differentiate the product of $(3x + 7)$ and $(4x^2 + 6x)$. Suppose we let $u = 3x + 7$ and $v = (4x^2 + 6x)$. Then it is easy to see that $\frac{du}{dx} = 3$, $\frac{dv}{dx} = 8x + 6$. Hence

$$\frac{d}{dx}[uv] = u\frac{dv}{dx} + v\frac{du}{dx} = (3x + 7)(8x + 6) + (4x^2 + 6x)(3).$$

Try this problem:

What is $\frac{d}{dx}[(2x + 3)(x^5)]$?

Answer: _____

Go to 193 for the solution.

193

$$\frac{d}{dx}[(2x + 3)(x^5)] = (2x + 3)(5x^4) + (x^5)(2)$$

The method for obtaining this is like that shown in frame 192. You can use the rule in frame 180 for differentiating x^n in order to find $\frac{d}{dx}x^5 = 5x^4$.

Go to 194.

194

In this frame we are going to learn the rule for finding the derivative of a "function of a function." Suppose w is a variable that depends on u, and u depends on x. Then w also depends on x, and the following rule, which is proved in Appendix A7, is very helpful.

$$\frac{dw}{dx} = \frac{dw}{du}\frac{du}{dx}$$

This rule is called the *chain rule*, because it links together derivatives with related variables. It is one of the most frequently used rules in differential calculus.

Here is an example: Suppose we want to differentiate $w = (x + x^2)^2$. This is a complicated function. It looks much simpler if we let $u = x + x^2$, in which case $w = u^2$, and $\frac{dw}{du} = 2u$. Then

$$\frac{dw}{dx} = \frac{dw}{du}\frac{du}{dx} = 2u\,\frac{du}{dx}.$$

We now substitute the value $u = x + x^2$, and $\frac{du}{dx} = 1 + 2x$, to obtain

$$\frac{dw}{dx} = 2(x + x^2)\,(1 + 2x)$$

(You can check that the chain rule gives the right answer in this case by multiplying out the expression for w, and then differentiating it. You will find that the answer is equivalent to $\frac{dw}{dx}$ found above.) In the next frame we will see some problems which can't be written in such a simple form and for which the chain rule is essential.

Go to 195.

195

Here are a few more examples of the use of the *chain rule*.

(1) Find $\dfrac{d}{dt} \sqrt{1 + t^2}$

Suppose we let $w = \sqrt{1 + t^2}$, and $u = 1 + t^2$, so that $w = \sqrt{u}$.

Then $\dfrac{dw}{dt} = \dfrac{dw}{du} \dfrac{du}{dt} = \dfrac{1}{2\sqrt{u}} (2t)$

$$= \frac{1}{2} \frac{1}{\sqrt{1 + t^2}} \, 2t = \frac{t}{\sqrt{1 + t^2}} \, .$$

Here we have used t as a variable, but of course it makes no difference what we call the variables.

(2) Let $v = \left[q^3 + \dfrac{1}{q} \right]^{-3}$; find $\dfrac{dv}{dq}$.

This problem can be simplified by letting $p = q^3 + \dfrac{1}{q}$ and $v = p^{-3}$. With these symbols the chain rule is

$$\frac{dv}{dq} = \frac{dv}{dp} \frac{dp}{dq} = -3p^{-4} \frac{dp}{dq} = -3p^{-4} \left[3q^2 - \frac{1}{q^2} \right]$$

$$= -3 \left[q^3 + \frac{1}{q} \right]^{-4} \left[3q^2 - \frac{1}{q^2} \right] .$$

The following example will not be explained, since you should be able to work it by inspection.

(3) $\dfrac{d}{dx} \left[1 + \dfrac{1}{x} \right]^2 = 2 \left[1 + \dfrac{1}{x} \right] \left[-\dfrac{1}{x^2} \right]$

Go to 196.

196

Now do the following problem:

Which expression correctly gives

$$\frac{d}{dx} (2x + 7x^2)^{-2}?$$

(a) $-2(2 + 14x)^{-3}$

(b) $-2(2 + 14x)^{-2}(2x + 7x^2)$

(c) $(2x + 7x^2)^{-3}(2 + 14x)$

(d) $-2(2x + 7x^2)^{-3}(2 + 14x)$

The correct answer is

$$\boxed{a \mid b \mid c \mid d}$$

If right, go to 199.
Otherwise, go to 197.

197

Here is how to work the problem in 196. Suppose we let $w = u^{-2}$, and $u = (2x + 7x^2)$.

Then $\dfrac{du}{dx} = (2 + 14x)$.

Hence

$$\frac{dw}{dx} = \frac{dw}{du}\frac{du}{dx} = \frac{d}{du}(u^{-2})\frac{du}{dx}$$

$$= -2u^{-3}\frac{du}{dx} = -2(2x + 7x^2)^{-3}(2 + 14x).$$

Try this problem:

Find $\dfrac{dw}{ds}$, where $w = 12q^4 + 7q$, and $q = s^2 + 4$.

$$\frac{dw}{ds} = \underline{\hspace{6cm}}.$$

For the solution, go to 198.

198

   The problem in frame 197 can be solved by using the chain
rule:

$\dfrac{dw}{ds} = \dfrac{dw}{dq} \dfrac{dq}{ds}$. We are given that $w = 12q^4 + 7q$ and $q = s^2 + 4$, so

$\dfrac{dw}{dq} = 48q^3 + 7$, and $\dfrac{dq}{ds} = 2s$.

Substituting these, we have

$$\dfrac{dw}{ds} = [48q^3 + 7][2s] = [48(s^2 + 4)^3 + 7][2s].$$

   If you wrote this result, go on to 199. If you made a mistake
you should study the last few frames to make sure you understand the application of the chain rule. Don't be confused by
the names of variables. Then go to 199.

199

   The next useful rule for differentiation is one you may be
able to work out for yourself by applying the chain rule.

   The problem is to find $\dfrac{d}{dx}\left(\dfrac{1}{v}\right)$ in terms of v and $\dfrac{dv}{dx}$, where v
depends on x. Which of the following answers correctly gives
$\dfrac{d}{dx}\left(\dfrac{1}{v}\right)$?

$$\left[-\frac{1}{v^2}\frac{dv}{dx} \;\middle|\; 1/\frac{dv}{dx} \;\middle|\; \frac{dx}{dv} \;\middle|\; -\frac{dv}{dx} \;\middle|\; \text{none of these} \right]$$

If right, go to 201.
If wrong, go to 200.

Answer: (196) *d*

200

To find $\dfrac{d}{dx}\left[\dfrac{1}{v}\right]$ we apply the chain rule in the following way.

Suppose we let $w = \dfrac{1}{v} = v^{-1}$

$$\frac{dw}{dx} = \frac{dw}{dv}\frac{dv}{dx}, \text{ but } \frac{dw}{dv} = \frac{d}{dv}v^{-1} = -\frac{1}{v^2}, \text{ so}$$

$$\frac{d}{dx}\left[\frac{1}{v}\right] = -\frac{1}{v^2}\frac{dv}{dx}.$$

Go to 201.

201

Now, by combining the result of the last frame with what you have learned previously, you should be able to derive an expression for the derivative of the quotient of two functions. This is an extremely important relation. Try to work it out for yourself.

Find $\dfrac{d}{dx}\left[\dfrac{u}{v}\right]$ in terms of u, v, $\dfrac{du}{dx}$, $\dfrac{dv}{dx}$.

$$\frac{d}{dx}\left[\frac{u}{v}\right] =$$

_____

To check your answer, go to 202.

Answer: (199) $-\dfrac{1}{v^2}\dfrac{dv}{dx}$

202

You should have obtained the following rule (though possibly arranged differently):

$$\frac{d}{dx}\left[\frac{u}{v}\right] = \frac{v\,\dfrac{du}{dx} - u\,\dfrac{dv}{dx}}{v^2}.$$

If you wrote this or an equivalent statement, go on to 203. Otherwise, study the derivation below.

If we let $p = \dfrac{1}{v}$, then our derivative is that of the product of two variables.

$$\frac{d}{dx}\left[\frac{u}{v}\right] = \frac{d}{dx}\,[up] = u\,\frac{dp}{dx} + p\,\frac{du}{dx}.$$

Now $\dfrac{dp}{dx} = \dfrac{dp}{dv}\dfrac{dv}{dx} = -\dfrac{1}{v^2}\dfrac{dv}{dx}$, as in frame 200, so

$$\frac{d}{dx}\left[\frac{u}{v}\right] = -\frac{u}{v^2}\frac{dv}{dx} + \frac{1}{v}\frac{du}{dx} = \frac{v\,\dfrac{du}{dx} - u\,\dfrac{dv}{dx}}{v^2}.$$

Go to 203.

203

Solve the following problem:

$$\frac{d}{dx}\left[\frac{1 + x}{x^2}\right] = $$

_____.   *To see the correct answer, go to 204.*

204

The answer to the problem in 203 is

$$\frac{d}{dx}\left[\frac{1+x}{x^2}\right] = -\frac{2}{x^3} - \frac{1}{x^2}.$$

If right, go to 206.

If wrong, you should go to 205 for help.

205

Let $u = 1 + x$, $v = x^2$. Then $\frac{du}{dx} = 1$, $\frac{dv}{dx} = 2x$

$$\frac{d}{dx}\left[\frac{u}{v}\right] - \frac{v\frac{du}{dx} - u\frac{dv}{dx}}{v^2}$$

$$\frac{d}{dx}\left[\frac{u}{v}\right] = \frac{x^2 - (1+x)(2x)}{x^4} = \frac{1}{x^2} - \frac{2}{x^3}(1+x)$$

$$= \frac{-2}{x^3} - \frac{1}{x^2}.$$

Go to 206.

206

   Before going on to new material, let's summarize all the
rules for differentiation we have used so far. *You* fill in the
blank. *a* and *n* are constants, *u* and *v* are variables that depend
on *x*, *w* depends on *u*, which in turn depends on *x*.

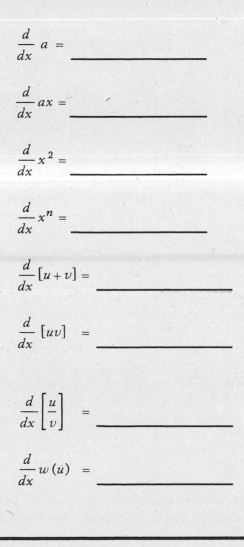

$$\frac{d}{dx} a = \underline{\hspace{3cm}}$$

$$\frac{d}{dx} ax = \underline{\hspace{3cm}}$$

$$\frac{d}{dx} x^2 = \underline{\hspace{3cm}}$$

$$\frac{d}{dx} x^n = \underline{\hspace{3cm}}$$

$$\frac{d}{dx} [u + v] = \underline{\hspace{3cm}}$$

$$\frac{d}{dx} [uv] = \underline{\hspace{3cm}}$$

$$\frac{d}{dx} \left[\frac{u}{v} \right] = \underline{\hspace{3cm}}$$

$$\frac{d}{dx} w(u) = \underline{\hspace{3cm}}$$

Go to 207.

207

Here are the correct answers, which you should have obtained with no trouble. The frame in which the relation was introduced is shown in parentheses.

$$\frac{d}{dx}\, a = 0 \tag{172}$$

$$\frac{d}{dx}\, ax = a \tag{174}$$

$$\frac{d}{dx}\, x^2 = 2x \tag{176}$$

$$\frac{d}{dx}\, x^n = nx^{n-1} \tag{180}$$

$$\frac{d}{dx}\, [u + v] = \frac{du}{dx} + \frac{dv}{dx} \tag{186}$$

$$\frac{d}{dx}\, [uv] = u\,\frac{dv}{dx} + v\,\frac{du}{dx} \tag{189}$$

$$\frac{d}{dx}\left[\frac{u}{v}\right] = \frac{v\,\dfrac{du}{dx} - u\,\dfrac{dv}{dx}}{v^2} \tag{202}$$

$$\frac{d}{dx}\, w\,(u) = \frac{dw}{du}\frac{du}{dx} \tag{194}$$

If you would like some more practice on problems similar to those in Sections 5 and 6, see review problems 34 through 38.

*Go to the next section,
frame 208.*

Section 7.   DIFFERENTIATING TRIGONOMETRIC FUNCTIONS

208

Trigonometric functions occur in so many applications that it is useful to know their derivatives. For instance, we would like to know $\dfrac{d}{d\theta} \sin \theta$. By definition,

$$\frac{d}{d\theta} \sin \theta = \lim_{\Delta\theta \to 0} \frac{\sin (\theta + \Delta\theta) - \sin \theta}{\Delta\theta}$$

It is not at all obvious how to evaluate this expression, so let's take another approach for a minute and try to guess *geometrically* what the result should be by looking at a plot of sin θ.

Here is a plot of sin θ vs θ. θ is the angle measured in radians, but for reference, a few of the angles are shown in degrees.

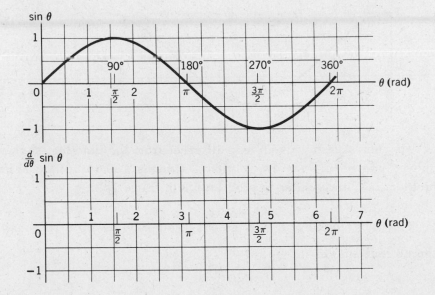

Draw a sketch of $\dfrac{d}{d\theta} \sin \theta$ in the space provided, and then, to check your drawing go to 209.

209

Here are drawings of $\sin \theta$ and $\dfrac{d}{d\theta} \sin \theta$. Note that where

the slope of $\sin \theta$ is greatest, at 0 and 2π, $\dfrac{d}{d\theta} \sin \theta$ has its great-

est value, and that where the slope is 0, at $\theta = \dfrac{\pi}{2}$ and $\dfrac{3\pi}{2}$, $\dfrac{d}{d\theta} \sin \theta$
is 0.

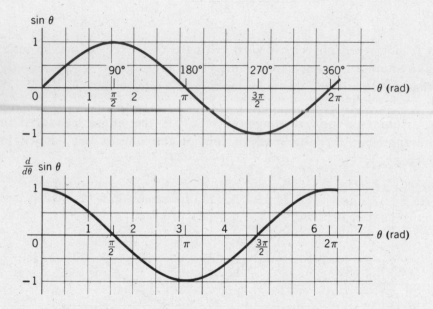

(If your sketch looked very different from the drawing shown
above, you should review Section 4 (frames 160 to 169). This
problem is quite similar to problem (c) in frame 168).

Now, by looking at the graphs, you may be able to guess
the correct answer for $\dfrac{d}{d\theta} \sin \theta$. Can you?

$$\frac{d}{d\theta} \sin \theta = \underline{\hspace{5cm}}.$$

*Go to frame 210 to see if
your answer is right.*

210

Here is the rule:

$$\frac{d}{d\theta} \sin \theta = \cos \theta.$$

Congratulations if you guessed this result in the last frame! If you arrived at some other result, you should study the drawings in frame 209 and compare the second one with the plot of $\cos \theta$ shown below (as well as in frame 65).

Formal proof that $\dfrac{d}{d\theta} \sin \theta = \cos \theta$ is given in Appendix A5.

It is important to realize that this relation is only true when angle is measured in radians, and this is why the radian is such a useful unit.

Let's try to guess the result for $\dfrac{d}{d\theta} \cos \theta$ from a plot of $\cos \theta$.

Draw a sketch of $\dfrac{d}{d\theta} \cos \theta$ in the space provided, and make a guess at the result.

$$\frac{d}{d\theta} \cos \theta = \underline{\hspace{5cm}}$$

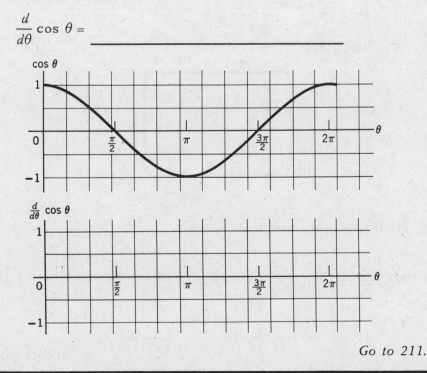

Go to 211.

211

Here are plots of $\cos \theta$ and $\dfrac{d}{d\theta} \cos \theta$. The result is $\dfrac{d}{d\theta} \cos \theta$ $= - \sin \theta$, as should seem reasonable from the graph. This relation also is formally proved in Appendix A5.

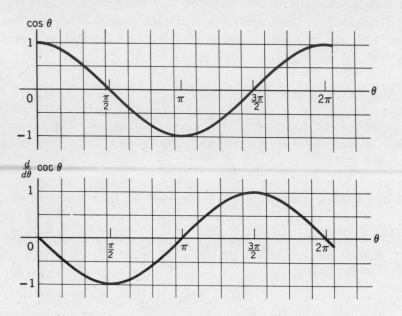

To summarize:

$$\frac{d}{d\theta} \sin \theta = \cos \theta$$

$$\frac{d}{d\theta} \cos \theta = - \sin \theta$$

Using these results, find $\dfrac{d}{d\theta} \tan \theta$.

(Hint: use $\tan \theta = \dfrac{\sin \theta}{\cos \theta}$ and apply the rule in frame 202.)

$\dfrac{d}{d\theta} \tan \theta = $ _____

Go to 212

212

Using the hints in frame 211 we have

$$\frac{d}{d\theta} \tan \theta = \frac{d}{d\theta}\left[\frac{\sin \theta}{\cos \theta}\right]$$

$$= \frac{\cos \theta \dfrac{d}{d\theta} \sin \theta - \sin \theta \dfrac{d}{d\theta} \cos \theta}{\cos^2 \theta}$$

$$= \frac{\cos^2 \theta + \sin^2 \theta}{\cos^2 \theta} = \frac{1}{\cos^2 \theta} = \sec^2 \theta.$$

Now find the correct answer:

$$\frac{d}{d\theta} \sec \theta = \boxed{\sec \theta \tan \theta \mid - \sec \theta \tan \theta \mid \sec \theta}.$$

If right, go to 214.
If wrong, go to 213.

213

Did you forget what you learned in the last section or have you not remembered the definitions of the trigonometric functions? You should have seen with the aid of frame 200 that

$$\frac{d}{d\theta} \sec \theta = \frac{d}{d\theta}\frac{1}{\cos \theta} = -\frac{1}{\cos^2 \theta}\frac{d \cos \theta}{d\theta}$$

$$= +\frac{1}{\cos^2 \theta} \sin \theta = \frac{\tan \theta}{\cos \theta}$$

$$= \sec \theta \tan \theta.$$

(All three of these expressions are equally acceptable.)

Go to 214.

214

Choose the correct answer:

$$\frac{d}{d\theta} (\sin \theta)^2 = \boxed{\sin \theta \mid 2 \cos \theta \mid \cos \theta^2 \mid 2 \sin \theta \cos \theta}$$

If right, go to 216.
If wrong, go to 215.

215

You could have analyzed the problem as follows:

Suppose we let $u(\theta) = \sin \theta$.

Then $\dfrac{du}{d\theta} = \cos \theta$, and

$$\frac{d}{d\theta}(\sin \theta)^2 = \frac{d}{d\theta} u^2 = \frac{d}{du}(u^2)\frac{du}{d\theta}$$

$$= 2u\frac{du}{d\theta} = 2\sin \theta \cos \theta.$$

Where did you go wrong?   Find your error and be sure you understand it. Then —

go to 216.

216

Which of the following is $\dfrac{d}{d\theta}\cos(\theta^3)$?

$$\left[\cos \theta \sin(\theta^3) \mid -3\,\theta^2 \sin(\theta^3) \mid 3\cos^2(\theta^3)\sin(\theta^3) \mid 3\cos^2\theta\right]$$

If right, skip on to frame 220.
If wrong, go to frame 217.

217

Did you forget how to use the *chain rule* to differentiate a function of a function? We can think of $\cos(\theta^3)$ as a function of a function. Suppose we write it this way:

$$w = \cos u, \quad u = \theta^3. \text{ Then}$$

$$\frac{dw}{d\theta} = \frac{dw}{du}\frac{du}{d\theta}$$

$$\frac{dw}{du} = -\sin u = -\sin(\theta^3), \qquad \frac{du}{d\theta} = 3\theta^2$$

so $\dfrac{d}{d\theta}\cos(\theta^3) = -3\,\theta^2 \sin \theta^3$

Go to 218.

Answers: (212) $\sec \theta \tan \theta$; (214) $2\sin \theta \cos \theta$

218

If ω (Greek letter omega) is a constant, which expression correctly gives $\dfrac{d}{dt}\sin\omega t$?

$\begin{bmatrix}\cos\omega t \mid \omega\cos\omega t \mid \sin\omega t \mid \text{none of these}\end{bmatrix}$

If right, go to frame 220.
Otherwise, go to 219.

219

To solve problem 218, let $w = \sin u,\ u = \omega t.\ \dfrac{dw}{dt} = \dfrac{dw}{du}\dfrac{du}{dt} =$

$\cos u \times \dfrac{d}{dt}(\omega t) = \omega\cos\omega t.$

Go to frame 220.

220

Before you go on to the next section, let's state once more the important relations we have introduced in this section:

$$\boxed{\begin{aligned} \frac{d}{d\theta}\sin\theta &= \cos\theta \\[2mm] \frac{d}{d\theta}\cos\theta &= -\sin\theta \end{aligned}}$$

There are two more functions which are so common that we ought to know their derivatives by heart: logarithmic and exponential, and to learn about them

go to the next section,
frame 221.

Answer: (216) $-3\,\theta^2\sin(\theta^3)$

Section 8. DIFFERENTIATION OF LOGARITHMS
AND EXPONENTIALS

221

We are about to find the derivative of a logarithm. If you feel shaky about your knowledge of logarithms, you should review Section 5 of Chapter I (p. 42) before going on to the next frame.

Go to 222.

222

You are already familiar with logarithms to the base 10. We can equally well manipulate logs with any base (as in frame 95). For instance we could use the number 2 or π as a base. However, for reasons that will soon become apparent, it is particularly convenient to use as a base the number e which has the value

$$e = 2.71828 \ldots\ldots$$

(Like π which has the value $3.14159 \ldots\ldots$, e is also an irrational number whose value can be calculated as accurately as we please. Appendix A8 shows how to do this.)

Go to 223.

Answer: (218) $\omega \cos \omega t$

223

Since we are going to use logarithms to the base e often in this section, let's devote a few frames to becoming familiar with them.

To begin with, we will find a rule for finding $\log_e x$ from $\log_{10} x$. The rule, which is proved below, is

$$\log_e x = \frac{\log_{10} x}{\log_{10} e} = 2.303 \log_{10} x.$$

This shows that the process of changing logarithms from the base 10 to the base e simply involves multiplication by a constant.

Here is how the rule is derived:

By definition of the logarithm

$$x = e^{\log_e x}.$$

Now take the logarithm to the base 10 of both sides of the equation.

$$\log_{10} x = \log_{10} (e^{\log_e x})$$

The right hand side can be simplified by using the rule $\log(x^n) = n \log x$, when n is any number. Thus we obtain,

$$\log_{10} x = \log_e x \times \log_{10} e.$$

or

$$\log_e x = \frac{\log_{10} x}{\log_{10} e}$$

but

$$\frac{1}{\log_{10} e} = \frac{1}{\log_{10} 2.718} = \frac{1}{0.4343} = 2.303 \ldots.$$

hence

$$\log_e x = 2.303 \log_{10} x.$$

Go to 224.

224

To see whether you have caught on, try the following problem.

From tables, you can verify that

$$\log_{10} 37 = 1.57.$$

Which one of the following is closest to $\log_e 37$?

$\boxed{1.57/e \mid 3.61 \mid 15.7 \mid 0.68}$

If right, go to 226.
If wrong, go to 225.

225

The method described in frame 223 leads directly to the solution:

$$\log_e 37 = \frac{1}{\log_{10} e} \log_{10} 37 = 2.303 \log_{10} 37$$

$$= 2.303 \times 1.57$$

$$= 3.61.$$

(You should have been able to pick the correct answer without doing the arithmetic exactly since all the other answers were far off.)

Go to 226.

226

Logarithms to the base e are called *natural logarithms*. They are so important that they are given a special symbol, ln x.

$$\boxed{\ln x = \log_e x}$$

Go to 227.

227

Here is a table showing ln x for a few values of x.

x	ln x		x	ln x
1	0.000		30	3.40
2	0.69		100	4.61
e	1.00		300	5.70
3	1.10		1000	6.91
10	2.30		3000	8.01

Using the table and the rules for manipulating logarithms you should be able to find the answer which is most nearly correct for each of the following questions:

$$\ln 6 - \boxed{2.2 \mid 3.1 \mid 6/e \mid 1.79}$$

$$\ln (\sqrt{10}) = \boxed{1.15 \mid 2.35 \mid 2.25 \mid 1.10}$$

$$\ln (300^3) = \boxed{126 \mid 185 \mid 17.10 \mid 3.41}$$

*If all your answers are correct,
go to 229.
If you made any mistakes,
go to 228.*

228

Before looking at the solution, make sure you are familiar with the rules for manipulating logarithms given in frame 91. Since these rules apply to logarithms of any base, they hold true for ln x.

$$\ln 6 = \ln (2 \times 3) = \ln 2 + \ln 3 = 0.69 + 1.10 = 1.79$$

$$\ln (\sqrt{10}) = \ln (10^{1/2}) = \frac{1}{2} \ln 10 = \frac{1}{2} \times 2.30 = 1.15$$

$$\ln (300^3) = 3 \ln 300 = 3 \times 5.70 = 17.10$$

Go on to 229.

Answer: (224) 3.61

229

Here is a plot of ln x in terms of x:

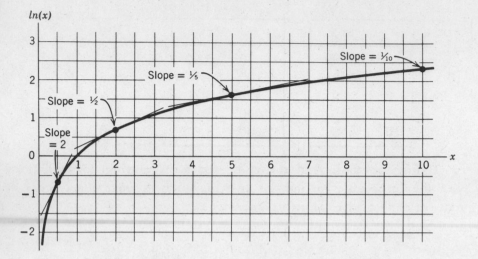

You should be able to see the qualitative features of $\dfrac{d}{dx}$ ln x by looking at the graph. For small values of x the derivative is large and for large values of x the derivative approaches 0. In the figure above tangents are shown at a few points, and their slopes are listed in this table.

x	slope
1/2	2
2	1/2
5	1/5
10	1/10

Perhaps you can guess the formula for $\dfrac{d}{dx}$ ln x. Try to fill in the blank.

$$\frac{d}{dx} \ln x = \underline{\hspace{4cm}}.$$

To learn the correct expression, go to 230.

Answers: (227) 1.79, 1.15, 17.10

230

Here is the formula for the derivative of a natural logarithm:

$$\frac{d}{dx} \ln x = \frac{1}{x}$$

If you did not guess this result you should check that it agrees with the numerical values in the table in frame 229.

The reason that e is so useful as a base for logarithms is that it leads to such a simple expression. This relation is derived in Appendix A9. It is so important that you should commit it to memory.

Go to 231.

231

Try this problem: Which of the following gives $\frac{d}{dx}[\ln (x^2)]$?

$$\left[2 \ln x \mid \frac{2}{x} \mid \frac{1}{x^2} \mid \frac{2}{x^2} \mid \frac{2}{x} \ln x \right]$$

If right, go to 234.
Otherwise, go to 232.

232

The solution of this problem is quite straightforward. We could make use of the chain rule. However, let's solve it another way.

Since $\ln (x^2) = 2 \ln x$, $\quad \dfrac{d}{dx} \ln (x^2) = \dfrac{d}{dx} 2 \ln x = \dfrac{2}{x}.$

You should be able to do this one —

$$\frac{d}{dx} (\ln x)^2 = \left[2 \ln x \mid \frac{2 \ln x}{x} \mid \frac{2\pi}{\ln x} \mid \text{none of these} \right]$$

If right, go to 234.
Otherwise, go to 233.

233

$$\frac{d}{dx} (\ln x)^2 = 2 \ln x \frac{d}{dx} \ln x = \frac{2 \ln x}{x}$$

Go to 234.

234

Although we know how to differentiate logs to the base e, we don't yet have a rule for differentiating logs to the base 10. However, it is easy to derive the rule. From frame 223 we have

$$\log_{10} x = \ln x \, \log_{10} e.$$

But $\log_{10} e$ is a constant, hence

$$\frac{d}{dx} \log_{10} x = \left[\frac{d}{dx} \ln x \right] \log_{10} e = \frac{1}{x} \log_{10} e = \frac{0.4343}{x}$$

Go to 235.

235

Let's put some of this to practice. Work out the correct answers:

(a) $\dfrac{d \ln r}{dr}$ = _____

(b) $\dfrac{d \ln 5z}{dz}$ = _____

For the correct answers,
go to 236.

236

The correct answers are —

(a) $\dfrac{1}{r}$ (b) $\dfrac{1}{z}$

If you got both of these, you are doing fine, so you may skip ahead to frame 238. If you missed either one

go to frame 237.

Answers: (231) $\dfrac{2}{x}$;   (232) $\dfrac{2 \ln x}{x}$

237

(a) You should immediately have seen that $\dfrac{d \ln r}{dr} = \dfrac{1}{r}$ since it makes no difference whether the variable is called r or x.

(b) The simplest way to find $\dfrac{d}{dz} \ln 5z$ is to recall that $\ln 5z = \ln 5 + \ln z$. Hence,

$$\frac{d}{dz} \ln 5z = \frac{d}{dz} \ln 5 + \frac{d}{dz} \ln z = 0 + \frac{1}{z} = \frac{1}{z}.$$

Go to 238.

238

Another form of function we would like to differentiate is

$$y = a^x \quad (a \text{ is a constant}).$$

We can do so with the aid of what we have just learned about logarithms. Take the natural logarithm of both sides of the above expression:

$$\ln y = \ln (a^x) = x \ln a.$$

Now differentiate both sides of this equation with respect to x

$$\frac{d}{dx} (\ln y) = \frac{dx}{dx} \ln a$$

$$\frac{1}{y} \frac{dy}{dx} = \ln a$$

$$\frac{dy}{dx} = y \ln a = a^x \ln a.$$

Thus $\dfrac{d}{dx} (a^x) = a^x \ln a.$

Go to 239.

239

The preceding frame gave the important result

$$\frac{d(a^x)}{dx} = a^x \ln a$$

A particularly simple case occurs when $a = e$. Since $\ln e = 1$ as a result of the definition of the natural logarithm,

$$\boxed{\frac{de^x}{dx} = e^x}$$

With the above can you write the values for the following?

(a) $\dfrac{de^{cx}}{dx} =$ _____

(b) $\dfrac{de^{-x}}{dx} =$ _____

See 240 for the
correct answers.

240

You should have written

(a) $\dfrac{de^{cx}}{dx} = ce^{cx}$

and

(b) $\dfrac{de^{-x}}{dx} = -e^{-x}.$

If you did both of these correctly, go to 241. Otherwise, continue here.

The result (a) is obtained by letting $u = cx$ and following the usual procedure for a function of a function (i.e., using the chain rule, frame 194 p. 110). Thus

$$\frac{de^{cx}}{dx} = \frac{de^u}{du}\frac{du}{dx} = e^u c = ce^{cx}.$$

The result (b) is a special case of (a) with $c = -1$.

Go to 241.

241

If $z = \dfrac{1}{\ln (x)}$, what is $\dfrac{dz}{dx}$?

Encircle the correct answer.

$$\left[\dfrac{1}{x \ln (x)} \quad \Big| \quad \dfrac{-x}{(\ln x)^2} \quad \Big| \quad \dfrac{-1}{x (\ln x)^2} \quad \Big| \quad \dfrac{\ln x}{x^2} \right]$$

If right, go to 243.
Otherwise, go to 242.

242

We can find the derivative of $\dfrac{1}{\ln (x)}$ by using the chain rule.

Let $u = \ln (x)$. Then

$$\dfrac{d}{dx} \dfrac{1}{\ln (x)} = \dfrac{d}{dx} \left[\dfrac{1}{u} \right] = \dfrac{du}{du} \overset{1}{} \dfrac{du}{dx} = -\dfrac{1}{u^2} \dfrac{1}{x}$$

$$= -\dfrac{1}{x (\ln x)^2}$$

Go to 243.

243

Since a number of relations have been introduced in this section, you may want to give them a quick review before going on. Here is a list of the relations: The most important ones are in boxes.

$$e = 2.71828 \ldots$$

$$\ln x = \log_e x$$

$$\ln x = 2.303 \log_{10} x$$

$$\boxed{\frac{d}{dx} \ln x = \frac{1}{x}}$$

$$\frac{d}{dx} (\log_{10} x) = \frac{0.4343}{x}$$

$$\frac{d}{dx} (a^x) = a^x \ln a$$

$$\boxed{\frac{d}{dx} e^x = e^x}$$

Go to 244.

244

We have learned how to differentiate the most useful common functions. The rest of this chapter will be spent on some special topics related to the use of derivatives. However, you may want a little more practice in differentiation before you go on. If so, see problems 34 through 58 on page 277. Whenever you are ready

go to Section 9, frame 245.

Answer: (241) $\dfrac{-1}{x (\ln x)^2}$

Section 9.   HIGHER ORDER DERIVATIVES

245

Suppose y depends on x and we have obtained the derivative $\frac{dy}{dx}$. If we next differentiate $\frac{dy}{dx}$ with respect to x, the result is called the *second derivative* of y with respect to x, and is written $\frac{d^2y}{dx^2}$.

Can you do the following problem?

If $y = 2x^3$, then $\frac{d^2y}{dx^2} = \left[\, 6x^2 \mid 12x \mid 0 \mid x^2 \mid x \,\right]$.

If right, go to 248.
If wrong, go to 246.

246

Here's how to do the problem in 245.

$$y = 2x^3$$

$$\frac{dy}{dx} = 6x^2$$

$$\frac{d^2y}{dx^2} = \frac{d}{dx}\left(\frac{dy}{dx}\right) = \frac{d}{dx}(6x^2) = 12x$$

Try this one:

$$y = x + \frac{1}{x}$$

$$\frac{d^2y}{dx^2} = \left[\, -\frac{1}{x^2} \mid \frac{1}{x} \mid +\frac{2}{x^3} \mid \text{none of these} \,\right].$$

If right, go to 248.
If wrong, go to 247.

247

Here is the solution to 246.

$$y = x + \frac{1}{x}$$

$$\frac{dy}{dx} = 1 - \frac{1}{x^2}$$

$$\frac{d^2y}{dx^2} = 0 - 1\left(\frac{-2}{x^3}\right) = \frac{2}{x^3}$$

Go to 248.

248

An example of a second derivative with which you may already be familiar is *acceleration*.

Velocity is the rate of change of position with respect to time.

$$v = \frac{dS}{dt}$$

Acceleration, a, is the rate of change of velocity with respect to time. Hence

$$a = \frac{dv}{dt}.$$

It follows then that

$$a = \frac{d}{dt}\left(\frac{dS}{dt}\right) = \frac{d^2S}{dt^2}.$$

Go to 249.

Answers: (245) $12x$;   (246) $\dfrac{2}{x^3}$

249

The position of a particle is given by

$$S = A \sin \omega t. \quad A \text{ and } \omega \text{ (omega) are constants.}$$

Find the acceleration.

Answer: $\boxed{0 \mid A\omega \cos \omega t \mid (A\omega \cos \omega t)^2 \mid - A\omega^2 \sin \omega t}$.

If right, go to 251.
If wrong, go to 250.

250

$$\text{Acceleration} = \frac{d^2 S}{dt^2} = \frac{d^2}{dt^2} (A \sin \omega t)$$

$$\frac{dS}{dt} = \frac{d}{dt} A \sin \omega t = A \omega \cos \omega t \quad \text{(See frame 218)}$$

$$\frac{d^2 S}{dt^2} = \frac{d}{dt} \left(\frac{dS}{dt} \right) = \frac{d}{dt} A \omega \cos \omega t = - A \omega^2 \sin \omega t.$$

Go to 251.

251

You see, there is really nothing new about a second derivative.

In fact, we can define derivatives of any order. $\frac{d^n f}{dx^n}$ is the n'th derivative of f with respect to x. Try this problem:

Suppose $y = x^4$. Then

$$\frac{d^4 y}{dx^4} = \boxed{x^{16} \mid 4x^4 \mid 0 \mid 64 \mid 4 \times 3 \times 2 \times 1}$$

Go to 252.

252

$$\frac{d^4}{dx^4}(x^4) = \frac{d}{dx}\left\{\frac{d}{dx}\left[\frac{d}{dx}\left(\frac{d}{dx} x^4\right)\right]\right\}$$

$$= \frac{d^3}{dx^3} 4x^3 = \frac{d^2}{dx^2} 4 \times 3 \, x^2 = \frac{d}{dx} 4 \times 3 \times 2 \, x$$

$$= 4 \times 3 \times 2 \times 1$$

We can easily generalize this result:

$$\frac{d^n}{dx^n} x^n = n \times (n-1) \times (n-2) \times \ldots \ldots 1$$

$$= n!$$

($n!$ is called n factorial and is $n \times (n-1) \times (n-2) \ldots \ldots 1$.)

For more practice on higher order derivatives, see problems 59 through 63 on p. 278.

Go on to Section 10,
frame 253.

Answer: (249) $-A\omega^2 \sin \omega t$

Section 10.   MAXIMA AND MINIMA

253

Now that we know how to differentiate simple functions, let's put our knowledge to use. Suppose we want to find the value of x and y at which

$$y = f(x)$$

has a minimum or a maximum value. By the end of this section we will know how to solve this problem.

Go to 254.

254

Here is the graph of a function. At which of the points indicated does y have a *minimum* value?

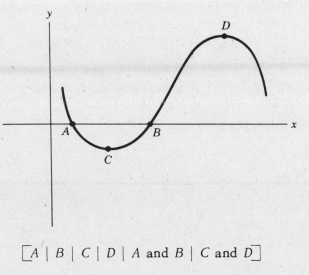

$$\boxed{A \mid B \mid C \mid D \mid A \text{ and } B \mid C \text{ and } D}$$

If correct, go to 256.
If wrong, go to 255.

Answer: (251) $4 \times 3 \times 2 \times 1$

255

The minimum value of y is at point C only, since y has its smallest value at point C, at least for the range of x and y shown.

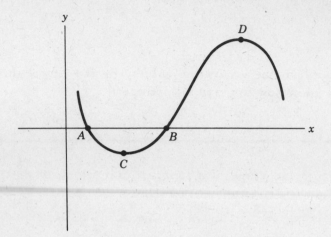

At A and B, y has the value 0, but this has nothing to do with whether or not it has a minimum value there.

Point D is a *maximum* value of y.

Go to 256.

Answer: (254) C

256

    We have shown that point C corresponds to a minimum value of y, at least insofar as nearby values are concerned, and that D corresponds similarly to a maximum value.

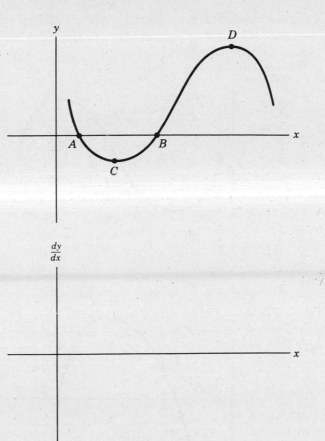

    There is an interesting relation between the points of maximum or minimum values of y and the value of the derivative at those points. To help see this, sketch a plot of the derivative of the function shown, using the space provided.

    To check your sketch,

go to 257.

257

Your sketch should look like the one shown here. Notice that at points C and D the derivative is 0, and that between C and D it is positive. Elsewhere it is negative.

If you did not obtain a sketch substantially like this, you should review Section 4 of this chapter before continuing.

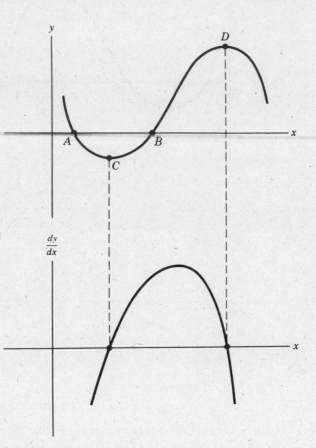

This simple example should be enough to convince you that:

If $f(x)$ has a maximum or minimum value for some value of x, then its derivative $\dfrac{df}{dx}$ is zero for that x.

One way to tell whether it is a maximum *or* a minimum is to plot a few neighboring points (although there is even a simpler method, as we shall soon see).   *Go to 258.*

258

Test yourself with this problem:

Find the value of x for which the following has a minimum value.

$$f(x) = x^2 + 6x$$

$$\left[-6 \mid -3 \mid 0 \mid +3 \mid \text{none of these} \right]$$

If right, go to 261.
If wrong, go to 259.

259

You should have solved the problem as follows:

The maximum or minimum occurs where x satisfies $\dfrac{d\,f(x)}{dx} = 0$.

$$f(x) = x^2 + 6x \qquad \frac{d\,f(x)}{dx} = 2x + 6$$

Thus the equation for the value of x at the maximum or minimum is

$$2x + 6 = 0, \quad \text{or} \quad x = -3.$$

Here is another problem:

For which values of x does the following $f(x)$ have a maximum or minimum value?

$$f(x) = 8x + \frac{2}{x}$$

$$\left[1/4 \mid -1/4 \mid -4 \mid 2 \text{ and } -4 \mid 1/2 \text{ and } -1/2 \right]$$

If you were right, go to 261.
*If you did not get the correct
answer, go to 260.*

260

The problem in frame 259 can be solved as follows:

At the position of maximum or minimum, $\dfrac{d}{dx} f(x) = 0$. Since

$$f(x) = 8x + \frac{2}{x}, \qquad \frac{d}{dx} f(x) = 8 - \frac{2}{x^2}.$$

The desired points are solutions of

$$8 - \frac{2}{x^2} = 0, \quad \text{or} \quad x^2 = \frac{2}{8} = \frac{1}{4}.$$

Thus at $x = +\dfrac{1}{2}$ and $x = -\dfrac{1}{2}$, $f(x)$ has a maximum or a minimum value. A plot of $f(x)$ is shown in the figure, and, as you can see, $x = -\dfrac{1}{2}$ yields a maximum, and $x = +\dfrac{1}{2}$ yields a minimum.

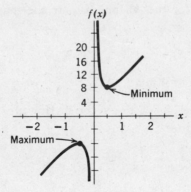

Incidentally, as you can see from the drawing, the minimum falls above the maximum. This should not be paradoxical, since we are talking about local minima or maxima — that is, the minimum or maximum value of a function in some small region.

Go to 261.

Answer: (258) −3;   (259) 1/2 and −1/2

261

We mentioned earlier that there is a simple method for finding whether $f(x)$ has a maximum or a minimum value when $\dfrac{d\,f(x)}{dx} = 0$. Let's find the method by drawing a few graphs.

Below are graphs of two functions. On the left, $f(x)$ has a maximum value in the region shown. On the right, $g(x)$ has a minimum value. In the spaces provided, draw rough sketches of the derivatives of $f(x)$ and $g(x)$.

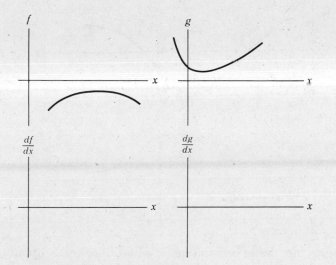

Now, let's repeat the process again. Make a rough sketch of the *second* derivative of each function (i.e., sketch the derivatives of the new functions you have just drawn).

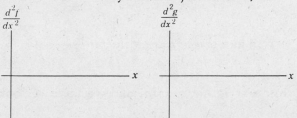

Perhaps from these sketches you can guess how to tell whether the function has a maximum or a minimum value when its derivative is 0. Whether you can or not,

go to 262.

262

The sketches should look approximately like this.

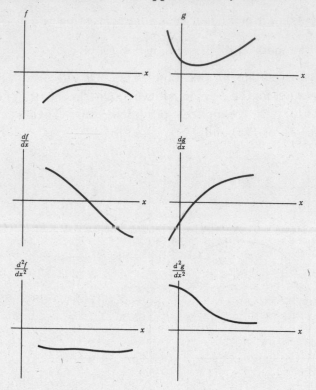

By studying these sketches, it should become apparent that wherever $\frac{df}{dx} = 0$,

$$f(x) \text{ has a } \textit{maximum} \text{ value if } \frac{d^2f}{dx^2} < 0 \text{ and}$$

$$f(x) \text{ has a } \textit{minimum} \text{ value if } \frac{d^2f}{dx^2} > 0.$$

(If $\frac{d^2f}{dx^2} = 0$, this test is not helpful and we have to look further.)

If you are not convinced yet, go back to Section 4 and sketch the second derivatives of any of the functions shown in frames 164, 166, or 168 [problems (c) or (d)]. This should convince you that the rule is reasonable. Whenever you are ready,

go to 263.

263

Here is one last problem to try before we go on to another subject. Consider $f(x) = e^{-x^2}$. Find the value of x for which $f(x)$ has a maximum or minimum value, and determine which it is.

Answer: _____

To check your answer, go to 264.

264

Let's solve the problem:

$f(x) = e^{-x^2}$. Using the chain rule, we find

$$\frac{df}{dx} = -2x \, e^{-x^2}.$$

Maximum or minimum occurs at x given by

$$-2x \, e^{-x^2} = 0, \quad \text{or} \quad x = 0.$$

Now we use the product rule (frame 189) to get

$$\frac{d^2f}{dx^2} = -2 \, e^{-x^2} + 4x^2 \, e^{-x^2} = (-2 + 4x^2) \, e^{-x^2}.$$

At $x = 0$, $\frac{d^2f}{dx^2} = (-2 + 4 \times 0) \times 1 = -2$. Since $\frac{d^2f}{dx^2}$ is negative

where $\frac{df}{dx} = 0$, at $x = 0$, $f(x)$ has a *maximum* value there.

A word of caution — in evaluating a derivative, say df/dx, at some value of x, $x = a$, you must always first differentiate $f(x)$ and then substitute $x = a$. If you reverse the procedure and first evaluate $f(a)$ and then try to differentiate it the result will simply be 0 since $f(a)$ is a constant. Similar care must be taken with higher order derivatives.

Go on to the next section, frame 265.

Section 11.   DIFFERENTIALS

265

So far we have denoted the derivative by the symbol $\frac{dy}{dx}$. Although this is a single symbol which stands for $\lim_{\Delta x \to 0} \frac{\Delta y}{\Delta x}$, the method of writing suggests that the derivative might be regarded as the ratio of two quantities, dy and dx. This turns out to be the case. The new quantities which we now introduce are called differentials, and they are defined in the next frame.

Go on to 266.

266

Suppose that x is an independent variable, and that $y = f(x)$. Then the *differential dx* of x is defined as equal to any increment, $x_2 - x_1$, where x_1 is the point of interest. The differential dx can be positive or negative, large or small, as we please. We see that dx, like x, can be regarded as an independent variable. The differential dy is defined by the following rule.

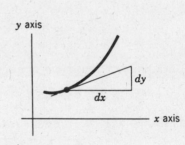

$$dy = \left[\frac{dy}{dx}\right] dx,$$

where $\left[\frac{dy}{dx}\right]$ is the derivative of y with respect to x.

Go to 267.

267

Although the meaning of the derivative, $\frac{dy}{dx}$, is $\lim_{\Delta x \to 0} \frac{\Delta y}{\Delta x}$, we can see from the preceding frame that it can now be interpreted as the ratio of the differentials dy and dx, where dx is any increment of x, and dy is defined by the rule $dy = \left[\frac{dy}{dx}\right] dx$.

Go to 268.

268

It is important not to confuse dy with Δy. As was pointed out in frame 136, Δy stands for $y_2 - y_1 = f(x_2) - f(x_1)$, where x_2 and x_1 are two given values of x. Both dx and Δx $(= x_2 - x_1)$ are arbitrary intervals. dx is called a *differential* of x, and Δx is called an increment of x, but their meanings are similar here. The diagram should show that dy and Δy are different quantities. Here we have set $dx = \Delta x$. The differential

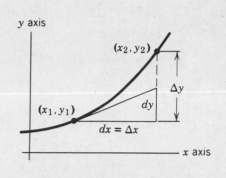

dy is then $\left[\dfrac{dy}{dx}\right] dx$, while the increment Δy is given by $y_2 - y_1$.

It is clear in this case that dy is not the same as Δy.

Go to 269.

269

Although dy and Δy are different, you can see from the figure that for sufficiently small dx (with $dx = \Delta x$) dy is very close to Δy. We can write this symbolically as

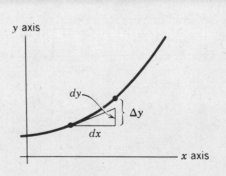

$$\lim_{dx=\Delta x \to 0} \frac{dy}{\Delta y} = 1$$

Hence, if we intend to take the limit where $dx \to 0$, dy may be substituted for Δy. Furthermore, even if we don't take the limit, dy is almost the same as Δy, provided dx is sufficiently small. We, therefore, often use dy and Δy interchangeably when it is understood that the limit will be taken or that the result may be an approximation.

Go to frame 270.

270

We can rewrite in differential form the various expressions for derivatives given earlier. Thus, if $y = x^n$,

$$dy = d(x^n) = \frac{d(x^n)}{dx} dx = nx^{n-1} dx.$$

Find the following:

$$d(\sin x) = \left[- \sin x \, dx \mid - \sin x \mid - \cos x \, dx \mid \cos x \, dx \right]$$

$$d(1/x) = \left[dx/x^2 \mid -dx/x^2 \mid -dx/x \right]$$

$$d(e^x) = \left[x \, e^x dx \mid dx \mid e^x \, dx \mid dx/e^x \right]$$

*If you missed any of these
go to 271.
Otherwise, go to 272.*

271

Here are the solutions to the problems in frame 270. The number of the frame in which each derivative is discussed is shown in parentheses.

$$d(\sin x) = \left[\frac{d \sin x}{dx} \right] dx = \cos x \, dx \quad \text{(frame 211)}$$

$$d(1/x) = \left[\frac{d}{dx} \left(\frac{1}{x} \right) \right] dx = - dx/x^2 \quad \text{(frame 180)}$$

$$d(e^x) = \left[\frac{d}{dx} e^x \right] dx = e^x \, dx \quad \text{(frame 239)}$$

Go to 272.

272

Here is an example of the use of a differential. The diagram shows the surface of a disc to which a thin rim has been added. Suppose we want an approximate value for the area of the rim.

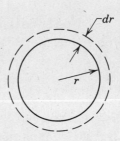

$$dA = \left[\frac{dA}{dr}\right] dr = \frac{d}{dr}(\pi r^2)\, dr = 2\pi r\, dr.$$

Go to 273.

273

The previous example can also be solved exactly by taking the difference of the two areas:

$$\Delta A = \pi (r + \Delta r)^2 - \pi r^2 = 2\pi r\, \Delta r + \pi \Delta r^2$$

When Δr is small compared to r we can neglect the last term and we see that

$$\Delta A = 2\pi r\, \Delta r.$$

If we let $\Delta r = dr$ and assume that they are both small then, as we know from frame 269,

$$dA = \Delta A = 2\pi r\, dr.$$

Here is a more intuitive argument for the results. Since the rim is thin, its area, dA, is the approximate length, $2\pi r$, multiplied by its width, dr. Hence,

$$dA = 2\pi r\, dr.$$

Go to 274.

Answers: (270) $\cos x\, dx, \quad -\dfrac{dx}{x^2}, \quad e^x\, dx$

274

Differentials are handy for remembering some important rules for differentiation. For instance, the chain rule

$$\frac{dw}{dx} = \frac{dw}{du}\,\frac{du}{dx}$$

is almost an identity if we treat dw, du and dx as differentials. Actually, it is not obvious that we can do so, since w and u both depend on a third quantity, x. Justification for using differentials to obtain the chain rule is given in Appendix A10.

Go to 275.

275

Here is another relation which is easy to remember with differentials, though the actual proof demands further explanation:

$$\frac{dx}{dy} = 1 \bigg/ \left[\frac{dy}{dx}\right]$$

This handy rule lets us reverse the role of dependent and independent variables, though it holds true only under certain conditions. If you want a further explanation, see Appendix A11.

*Otherwise, go to Section 12,
frame 276.*

Section 12.   A LITTLE REVIEW AND A FEW PROBLEMS

276

Let's end the chapter by reviewing some of the ideas introduced early in the chapter and by putting differential calculus to work in a few problems involving velocity.

Go to 277.

277

We hope you recall that the rate of change of position of a moving point with respect to time is called the velocity.

In other words, if position and time are related by a function S, in order to find the velocity, we _____ $S(t)$ with respect to _____ .

Go to 278.

278

You should have written:

In other words, if the position and time are related by a function S, in order to find the velocity, we *differentiate* $S(t)$ with respect to *time* (or t).

Go to 279.

279

Can you answer this problem?

The position of a particle along a straight line is given by the following expression:

$S = A \sin \omega t$. A and ω (omega) are constants.

Find the velocity of the particle.

$v =$ _____

For the answer, go to frame 280.

280

Your answer should have been

$$v = A\,\omega\,\cos\,\omega t.$$

If you got the right answer, skip on to 283. Otherwise, continue here.

The problem is to find the velocity, which is the rate of change of position with respect to time.

In this problem, the position is $S = A\,\sin\,\omega t$.

$$v = \frac{dS}{dt} = \frac{d}{dt}\,(A\,\sin\,\omega t) = A\,\omega\,\cos\,\omega t$$

(If you are not sure of the procedure here, see frame 218.)

Can you do this problem?

$$S = A\,\sin\,\omega t + B\,\cos\,2\,\omega t.\quad\text{Find }v.$$

$$v = \underline{\hspace{5cm}}$$

See frame 281 for the answer.

281

$$v = \frac{d}{dt}\,(A\,\sin\,\omega t + B\,\cos\,2\,\omega t)$$

$$= A\omega\,\cos\,\omega t - 2\,B\omega\,\sin\,2\omega t$$

If you wrote this, go to 283. If not review frame 219 and then continue here.

Try this problem: The position of a point is given by

$$S = A\,\sin\,\omega t\,\cos\,\omega t$$

Find its velocity.

$$v = \underline{\hspace{5cm}}$$

Go to 282 for the answer.

282

Here is how to solve problem 281.

$$v = \frac{dS}{dt} = \frac{d}{dt}(A \sin \omega t \cos \omega t)$$

$$= A \sin \omega t \frac{d}{dt}\cos \omega t + A\left(\frac{d}{dt}\sin \omega t\right)\cos \omega t$$

$$= -A\omega \sin^2 \omega t + A\omega \cos^2 \omega t$$

$$= A\omega (\cos^2 \omega t - \sin^2 \omega t)$$

As an alternative approach you might note that $\sin \omega t \cos \omega t$ $= \frac{1}{2}(\sin 2 \omega t)$. (See frame 71). Then, $v = \frac{d}{dt}\frac{A}{2}\sin 2 \omega t$. If you feel energetic, show that this procedure yields the same result as above.

Go to 283.

283

Suppose the height of a ball above the ground is given by $y = a + bt + ct^2$ where a, b, c, are constants. (Here we are using y rather than S to denote position. It makes no difference what we call our variable. This type of equation actually describes the height of a freely falling body.)

Find the velocity in the y direction.

$$v = \underline{\hspace{4cm}}$$

See 284 for the correct answer.

284

Here is how to do the problem in frame 283.

$$v = \frac{dy}{dt} = \frac{d}{dt}(a + bt + ct^2) = b + 2\,ct.$$

If you wrote the correct answer, go to 286. Otherwise, do the problem below.

Let $S = \dfrac{e}{t^2} + bt.$ (e and b are constants.)

Find the velocity.

$$v = \underline{\hspace{4cm}}$$

The answer is in
frame 285.

285

$$v = \frac{dS}{dt} = \frac{d}{dt}\left(\frac{e}{t^2} + bt\right) = \frac{-2e}{t^3} + b$$

If this problem gave you any difficulty you should review the beginning of this section before going on.

Otherwise, go to 286.

286

Here is a more difficult problem which you may enjoy. (If you don't feel in the mood, skip on to frame 288.)

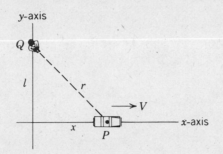

A car P moves along a road in the x direction with a constant velocity V. The problem is to find how fast it is moving away from a man standing at point Q, distance l from the road, as shown. In other words, if r is the distance between Q and P, find $\dfrac{dr}{dt}$.

(Hint — the chain rule is very useful here in the form $\dfrac{dr}{dt} = \dfrac{dr}{dx}\dfrac{dx}{dt}$.)

$$\frac{dr}{dt} = \underline{\hspace{4in}}.$$

Go to 287 after working this out

287

From the diagram in 286 you can see that

$$r^2 = x^2 + l^2, \quad r = (x^2 + l^2)^{1/2}$$

We must find $\dfrac{dr}{dt}$, and we can do so in the following way:

$$\frac{dr}{dt} = \frac{dr}{dx}\frac{dx}{dt} = \frac{d}{dx}(x^2 + l^2)^{1/2}\frac{dx}{dt}$$

$$= \frac{1}{2}\frac{2x}{(x^2 + l^2)^{1/2}} \times \frac{dx}{dt}$$

$$= V \times \frac{x}{(x^2 + l^2)^{1/2}}.$$

In the last step we have used $V = \dfrac{dx}{dt}$.

Go to Section 13,
frame 288.

Section 13.   CONCLUSION TO CHAPTER II.

288

This concludes your study of differential calculus, at least for the present. Perhaps all this material leaves you a little bewildered. If so — don't worry. The techniques will become much easier as you put them to use. Many of the proofs which were not given in the text are presented in Appendix A, and you may enjoy working through them.

In this chapter we have only discussed differentiation of functions of a single variable. It is not difficult to extend the ideas to functions of several variables. The technique for doing this is known as *partial differentiation*. If you are interested in the subject see Appendix B2. Another related topic, *implicit differentiation*, is explained in Appendix B3. This is a technique for differentiating functions which are not in the standard form $y = f(x)$. Appendix B4 shows how to differentiate the inverse trigonometric functions.

All the important results of this chapter are summarized in Chapter IV. You may wish to read that material now as a quick review. In addition, a list of important derivatives is presented in Table 1 at the back of the book.

Don't forget the review problems, page 275, if you want more practice.

Ready for more? Take a deep breath and go on to Chapter III.

CHAPTER III

INTEGRAL CALCULUS

We are now ready to tackle integral calculus. In this chapter you will learn

(1) how the indefinite integral is defined,

(2) the meaning of integration,

(3) how to find the area under any curve,

(4) how to evaluate definite integrals,

(5) how integral calculus can be applied to problems involving space and motion,

(6) how to use multiple integrals.

289

In this chapter we are going to learn about the second major branch of calculus — integral calculus. Integration is basically the opposite of differentiation; we are given the derivative and must find the function. That is, we are given a function f and are required to find another function F such that $\dfrac{dF(x)}{dx} = f(x)$.

Integration is useful in a great variety of applications. For instance, in some problems we obtain equations involving derivatives. Solving such equations requires integration. Integration also allows us to compute the area lying within different curves, as well as the volume of any solid whose limits can be expressed by equations. We will see a few examples of the last two applications later on in this chapter. To start off, go to 290.

290

Here is a simple problem to help you understand the meaning of integration.

Find a function F such that its derivative is x^2.

$F(x) = $ _____ .

Go to frame 291 for
the answer.

291

The correct answer is

$$F(x) = \frac{1}{3} x^3 + c.$$

c can be any constant; that is, c is an *arbitrary* constant. (Of course you can use any symbol you please for the constant. In this chapter c will always stand for a constant.) If you got this right and even remembered to put in the arbitrary constant you deserve double congratulations!

You can check that this function satisfies the requirement simply by differentiating it.

$$\frac{dF}{dx} = \frac{d}{dx}\left[\frac{1}{3} x^3\right] + \frac{d}{dx}(c) = x^2 + 0 = x^2$$

Try this problem: find a function G such that its derivative is $x + x^2$.

$$G(x) = \underline{\hspace{6cm}}$$

Go to 292 for the correct answer.

292

You should have written

$$G(x) = \frac{1}{2} x^2 + \frac{1}{3} x^3 + c, \text{ where } c \text{ is any constant.}$$

Again, you can verify that this is the correct solution by differentiating it.

$$\frac{dG}{dx} = \frac{d}{dx}\left[\frac{1}{2} x^2 + \frac{1}{3} x^3 + c\right] = x + x^2, \text{ as required.}$$

Go to 293.

293

Let's be a little more specific about what we are doing.

Suppose $\dfrac{dF(x)}{dx} = f(x)$.

Then $F(x)$ is called the *indefinite integral* of $f(x)$. This statement is written symbolically in the following form,

$$F(x) = \int f(x)\, dx.$$

The equation is read "$F(x)$ equals the indefinite integral of $f(x)$." The symbol $\int$ is called an integral sign. Sometimes we will write it $\int$. Don't confuse $\int$ with the letter f.

You will see shortly that this notation is really quite helpful, even though it may look mysterious now. Incidentally, the quantity dx which appears in the integral looks like a differential, and we will show shortly that it really is a differential. However, for the present, it is just part of the whole symbol.

The function which is integrated is called the *integrand*. In the above example, the integrand is $f(x)$.

Go to 294.

294

To make sure you understand the notation, fill in the blank below.

If $F(x) = \int f(x)\, dx$, then

$$\frac{dF(x)}{dx} = \underline{\hspace{5cm}}.$$

Go to 295.

295

The correct answer is

$$\frac{dF(x)}{dx} = f(x).$$

If you missed this, you should reread the last two frames and try to remember the new terms carefully.

                                                                     Go to 296.

296

Now we have learned that if

$$F(x) = \int f(x)\, dx$$

then

$$\frac{dF(x)}{dx} = f(x).$$

However, we can add any constant, c, to $F(x)$ and still meet the required condition, since

$$\frac{d}{dx}[F(x) + c] = \frac{dF(x)}{dx} + \frac{dc}{dx} = f(x).$$

Apparently then, if a constant is added to an indefinite integral of $f(x)$ the sum is also an indefinite integral of $f(x)$. Likewise we might expect that any two indefinite integrals of $f(x)$ can differ only by a constant. (This is proved in Appendix A12.) The label *indefinite* is used because c can have any value. Often, this label is dropped.

                                                                     Go to 297.

297

Let's pause for a moment to summarize.

If $f(x) = \dfrac{dF}{dx}$, then

F is the _____   _____ of ____ .

Go to 298 for
the answers.

298

If $f(x) = \dfrac{dF}{dx}$, then F is the *indefinite integral* of $f(x)$. If you
wrote something else, review the material starting at the beginning of this chapter. Otherwise, fill in the following:

Write an equation which is equivalent to asserting that $\dfrac{dy}{dx} = f(x)$.

$y =$ _____ .

Go to 299

299

You should have written

$y = \int f(x)\, dx.$

The quantity $\int f(x)\, dx$ is the *indefinite integral* of $f(x)$. Since
$\int f(x)\, dx$ depends on x, it defines a new function.

We can use any symbol to stand for $\int f(x)\, dx$: F, G, y, etc.
Generally, we will use the symbol F in this chapter.

Go to 300.

300

Try to find the indefinite integral of the following functions. c stands for a constant.

(a) $f(x) = \cos x$

$$\int \cos x \, dx = \left[\, \sin x + c \mid c \sin x \mid \cos x \mid \text{none of these} \,\right]$$

(b) $f(x) = \dfrac{1}{x^2}$

$$\int \frac{dx}{x^2} = \left[\, -\frac{c}{x^3} \mid -\frac{1}{x} + c \mid -\frac{1}{3x^3} \mid \text{none of these} \,\right]$$

If both correct,
go to Section 2, frame 302.
Otherwise, go to 301.

301

To check that the given answers meet the requirement we simply show that their derivatives yield the required functions.

(a) $\dfrac{d}{dx}(\sin x + c) = \cos x$, so $\sin x + c = \int \cos x \, dx$

(b) $\dfrac{d}{dx}\left(-\dfrac{1}{x} + c\right) = +\dfrac{1}{x^2}$, so $-\dfrac{1}{x} + c = \int \dfrac{1}{x^2} \, dx$

Go to Section 2, frame 302.

Section 2.   INTEGRATION

302

So far we have seen how to find indefinite integrals of a few specific algebraic functions. We did this by hunting for an expression which when differentiated gave the original function. In this section we will now see how this can be done more systematically.

Go to 303.

303

Since integration is the inverse of differentiation, for every differentiation formula in Chapter II, there is a corresponding integration formula here. Thus from Chapter II,

$$\frac{d \sin x}{dx} = \cos x,$$

so by the definition of the indefinite integral,

$$\int \cos x \, dx = \sin x + c.$$

Now you try one. What is

$$\int \sin x \, dx \; ?$$

Answer: $\left[\cos x + c \mid -\cos x + c \mid \sin x \cos x + c \mid \text{none of these} \right]$

Make sure you understand the correct answer (you can check the result by differentiation) and then go to 304.

304

Now try to find these integrals (for simplicity, the constant c has been omitted from the answers).

(a) $\displaystyle\int x^n \, dx = \left[\dfrac{1}{n} x^n \mid \dfrac{1}{n} x^{n+1} \mid \dfrac{1}{n+1} x^{n+1} \mid \dfrac{1}{n-1} x^n \right]$

(b) $\displaystyle\int e^x \, dx = \left[e^x \mid x \, e^x \mid \dfrac{1}{x} e^x \mid \text{none of these} \right]$

> *If you did both of these correctly you are doing fine and should skip to frame 306.*
>
> *If not, go to frame 305.*

Answers: (300) (a) $\sin x + c$, (b) $-\dfrac{1}{x} + c$

305

If you missed these due to a careless mistake and if you now understand the problem, correct your mistake and go on to frame 306. If not, review the definitions of the indefinite integral in the first section of this chapter and then continue here.

If $F = \int f(x)\, dx$

then

$$\frac{dF}{dx} = f(x).$$

Therefore if we want to find F we try to find an expression which when differentiated gives $f(x)$. Now the derivative of $\frac{x^{n+1}}{n+1}$ is given by

$$\frac{d}{dx}\left[\frac{x^{n+1}}{n+1}\right] = \frac{1}{n+1}\frac{dx^{n+1}}{dx} = \frac{1}{n+1}(n+1)x^n = x^n$$

by the formula for differentiating x^n in Chapter II.

Thus, including the integration constant c, we find $\int x^n\, dx = \frac{x^{n+1}}{n+1} + c.$

(Note, that this formula will not work for $n = -1$.)

Likewise, by Chapter II,

$$\frac{d}{dx}e^x = e^x$$

so

$$\int e^x\, dx = e^x + c$$

Go to 306.

Answers: (303) $-\cos x + c$; (304) (a) $\frac{1}{n+1}x^{n+1}$; (b) e^x

306

So far we have found integrals by looking for a function whose derivative is the integrand. Although this works well in many cases, especially after you have had some practice, it is helpful to have a list of some of the more important integrals. It is quite on the up and up to use such a list. If you make much use of calculus you will eventually know by sight most of the integrals listed, or at least be sufficiently familiar with them to make a good guess at the integral. You can always check your guess by differentiation.

A table of important integrals is given in the next frame. You can check the truth of any one of the equations

$$\int f(x)\, dx = F(x)$$

by confirming that

$$\frac{d\,F(x)}{dx} = f(x).$$

We will shortly use this method to verify some of the equations.

Go to 307.

307

LIST OF IMPORTANT INTEGRALS

In the following list of integrals u and v are variables that depend on x; w is a variable that depends on u which in turn depends on x; a and n are constants, and the arbitrary integration constants are omitted for simplicity.

(1) $\int a \, dx = ax$

(2) $\int a \, f(x) \, dx = a \int f(x) \, dx$

(3) $\int (u + v) \, dx = \int u \, dx + \int v \, dx$

(4) $\int x^n \, dx = \dfrac{x^{n+1}}{n + 1}$ $\qquad\qquad\qquad n \neq -1$

(5) $\int \dfrac{dx}{x} = \ln x$

(6) $\int e^x \, dx = e^x$

(7) $\int e^{ax} \, dx = e^{ax}/a$

(8) $\int b^{ax} \, dx = \dfrac{b^{ax}}{a \ln b}$

(9) $\int \ln x \, dx = x \ln x - x$

(10) $\int \sin x \, dx = -\cos x$

(11) $\int \cos x \, dx = \sin x$

(12) $\int \tan x \, dx = -\ln \cos x$

(13) $\int \cot x \, dx = \ln \sin x$

(14) $\int \sec x \, dx = \ln (\sec x + \tan x)$

(15) $\int \sin x \cos x \, dx = \dfrac{1}{2} \sin^2 x$

(16) $\int \dfrac{dx}{a^2 + x^2} = \dfrac{1}{a} \arctan \dfrac{x}{a}$

(17) $\int \dfrac{dx}{\sqrt{a^2 - x^2}} = \arcsin \dfrac{x}{a}$

(Continued on next page)

307 cont'd

LIST OF IMPORTANT INTEGRALS cont'd

(18) $\int \dfrac{dx}{\sqrt{x^2 \pm a^2}} = \ln\left[x + \sqrt{x^2 \pm a^2}\,\right]$

(19) $\int w(u)\,dx = \int \left[w(u)\dfrac{dx}{du}\right] du$

(20) $\int u\,dv = uv - \int v\,du$

For convenience this table is repeated as Table 2 near the back of the book (page 285).

Go to 308.

308

Let's see if you can check some of the formulas in the table. Show that integral formulas (9) and (15) are correct.

If you have proved the formulas to your satisfaction, go to 310.

If you want to see proofs of the formulas, go to 309.

309

To prove that $F(x) = \int f(x)\,dx$, we must show that $\dfrac{d\,F(x)}{dx} = f(x)$.

(9) $F(x) = x \ln x - x$, $f(x) = \ln x$

$$\frac{dF}{dx} = \frac{d}{dx}(x \ln x - x) = x\left(\frac{1}{x}\right) + \ln x - 1 = \ln x = f.$$

(15) $F(x) = \dfrac{1}{2}\sin^2 x$, $f(x) = \sin x \cos x = f.$

$$\frac{d}{dx}\frac{1}{2}\sin^2 x = \frac{1}{2}(2\sin x)\frac{d}{dx}\sin x = \sin x \cos x$$

Go to 310.

310

To prove formula 19 of frame 307, all we need to do is to show that the derivative with respect to x of the right side of the equation equals the integrand, $w(u)$. However by using the chain rule $\dfrac{dF}{dx} = \dfrac{dF}{du}\dfrac{du}{dx}$ we have:

$$\frac{d}{dx}\left[\int w(u)\frac{dx}{du}\,du\right] = \frac{d}{du}\left[\int w(u)\frac{dx}{du}\,du\right] \times \frac{du}{dx}$$

$$= \left[w(u)\frac{dx}{du}\right]\frac{du}{dx} = w(u).$$

In the last step we use $\dfrac{dx}{du} = 1/\dfrac{du}{dx}$ (proved in Appendix A11).

Go to 311.

311

We'll now discuss an important application of formula 19, which we have just proved.

When we write $\int f(x)\, dx$, the symbol dx looks like the symbol for the differential dx which we discussed in frame 266. On the other hand the dx in the integral sign is just part of the symbol and we have no reason to assume that it behaves like a differential. However, the formula we just proved indicates that dx can be treated as a differential in that we can replace dx by $\dfrac{dx}{du}\, du$, which is automatically implied by the differential notation.

Go to 312.

312

Our right to replace dx in the integral sign by $\dfrac{dx}{du}\, du$ gives us a powerful tool, since it implies that the integration is now performed with respect to du instead of dx. This substitution is often called a "change in variable". To see how useful it is, go to 313.

313

Here is an example to show how a *change* in *variable* can make an integral appear much simpler. Let us try to calculate $\int \sin 3x \, dx$. This integral has an unfamiliar form. However, it can be made to look like a familiar integral by a suitable change in variable.

$$\int \sin 3x \, dx = \frac{1}{3} \int 3 \sin 3x \, dx = \frac{1}{3} \int \sin 3x \, d(3x) = \frac{1}{3} \int \sin u \, du$$

where u is a new variable, $u = 3x$.

Then

$$\int \sin 3x \, dx = \frac{1}{3} \int \sin u \, du = \frac{-1}{3} [\cos u + c] = -\frac{1}{3} [\cos 3x + c].$$

(Our integral formulas in frame 307 are true no matter what the name of the independent variable, so if $\int \sin x \, dx = - \cos x + c$, then $\int \sin u \, du = - \cos u + c$.)

To see whether you have caught on, use this method to evaluate $\int \sin \frac{x}{2} \cos \frac{x}{2} \, dx$. (You may find formula 15 in frame 307 helpful.)

$$\int \sin \frac{x}{2} \cos \frac{x}{2} \, dx = \underline{\hspace{5cm}}$$

To check your answer,
go to 314.

314

Your answer should have been:

$$\int \sin \frac{x}{2} \cos \frac{x}{2}\, dx = \sin^2 \frac{x}{2} + c'. \qquad (c' = \text{any constant})$$

If you obtained this result, go right on to 316. Otherwise, continue here.

$$\int \sin \frac{x}{2} \cos \frac{x}{2}\, dx = 2 \int \sin \frac{x}{2} \cos \frac{x}{2}\, d\left(\frac{x}{2}\right) = 2 \int \sin u \cos u\; du,$$

where we have substituted $u = \frac{x}{2}$. But from formula 15 of frame 307 we have

$$\int \sin u \cos u\; du = \frac{1}{2} \sin^2 u + c = \frac{1}{2} \sin^2 \frac{x}{2} + c,$$

so

$$\int \sin \frac{x}{2} \cos \frac{x}{2}\, dx = 2\left(\frac{1}{2} \sin^2 \frac{x}{2} + c\right) = \sin^2 \frac{x}{2} + c'.$$

$(c' = 2c = \text{any constant}).$

Let's check this result:

$$\frac{d}{dx}\left[\sin^2 \frac{x}{2} + c'\right] = 2\left(\sin \frac{x}{2} \cos \frac{x}{2}\right)\left(\frac{1}{2}\right) = \sin \frac{x}{2} \cos \frac{x}{2}?$$

as required. (We have used the chain rule here.)

Try this problem: If a and b are constants, evaluate

$$\int \frac{dx}{a^2 + b^2 x^2} = \underline{\hspace{5cm}}.$$

You will find the table in frame 307 helpful in solving this.

Go to 315 for the solution.

315

Here are the steps to finding the required integral.

$$\int \frac{dx}{a^2 + b^2 x^2} = \frac{1}{b}\int \frac{d(bx)}{a^2 + b^2 x^2} = \frac{1}{b}\int \frac{du}{a^2 + u^2} \qquad (u = bx)$$

$$= \frac{1}{ab}\left[\arctan \frac{u}{a} + c\right] \qquad \text{(frame 307, formula 16)}$$

$$= \frac{1}{ab}\left[\arctan \frac{bx}{a} + c\right]$$

Go to 316.

316

We have just seen how to evaluate an integral by changing the variable from x to $u = ax$, where a is some constant. Often it is possible to simplify an integral by substituting still other quantities for the variable. Here is an example.

We want to evaluate $\displaystyle\int \frac{x\,dx}{x^2 + 4}$.

Suppose we let $u^2 = x^2 + 4$. Then $2u\,du = 2x\,dx$.

$$\int \frac{x\,dx}{x^2 + 4} = \int \frac{u\,du}{u^2} = \int \frac{du}{u} = \ln u + c = \ln \sqrt{x^2 + 4} + c$$

Try to use this method for evaluating the following integral:

$$\int \cos(2\theta + 5)\, d\theta.$$

Answer: _____

Go to 317
to check your answer.

317

We can find $\int \cos (2\theta + 5)\, d\theta$ in the following way.   Let $u = 2\theta + 5$, $du = 2d\theta$, then

$$\int \cos (2\theta + 5)\, d\theta = \frac{1}{2} \int \cos u\, du = \frac{1}{2}\, [\sin u + c]$$

$$= \frac{1}{2}\, [\sin (2\theta + 5) + c].$$

After you are accustomed to this procedure, you will often find that you can skip introducing u explicitly.   For instance, the last problem can be solved as follows:

$$\int \cos (2\theta + 5)\, d\theta = \frac{1}{2} \int \cos (2\theta + 5)\, d(2\theta + 5)$$

$$= \frac{1}{2}\, \sin (2\theta + 5) + c.$$

However, you should write out the intermediate steps until you are sure of yourself.

Go to 318.

318

Formula 20 in frame 307 is often helpful, and is known as "integration by parts". The proof is simple. Let u and v be any variables depending on x. Then from frame 189 on page 108

$$\frac{d}{dx}(uv) = u\frac{dv}{dx} + v\frac{du}{dx}.$$

Now integrate both sides of the equation with respect to x so, by frame 312,

$$\int \frac{d}{dx}(uv)\,dx = \int u\frac{dv}{dx}\,dx + \int v\frac{du}{dx}\,dx$$

$$\int d(uv) = \int u\,dv + \int v\,du.$$

But, $\int d(uv) = uv$, and, after transposing we have

$$\int u\,dv = uv - \int v\,du$$

Here is an example: Find $\int \theta \sin\theta\,d\theta$.

Let $u = \theta$, $dv = \sin\theta\,d\theta$. Then it is easy to see that $du = d\theta$, $v = -\cos\theta$.

Thus $\int \theta \sin\theta\,d\theta = \int u\,dv = uv - \int v\,du$

$$= -\theta\cos\theta - \int(-\cos\theta)\,d\theta$$

$$= -\theta\cos\theta + \sin\theta.$$

Go to 319.

319

Try to use integration by parts to find $\int xe^x\,dx$.

Answer: (constant omitted).

$\left[(x-1)e^x \mid xe^x \mid e^x \mid xe^x + x \mid \text{none of these}\right]$

If right, go to 321.
If you missed this, or want
to see how to
solve the problem, go to 320.

320

To find $\int xe^x \, dx$ using the formula for integration by parts, we can let $u = x$, $dv = e^x \, dx$, so that $du = dx$, $v = e^x$. Then,

$$\int xe^x \, dx = xe^x - \int e^x \, dx = xe^x - e^x$$

$$= (x - 1) \, e^x.$$

Go to 321.

321

Find the following integral using the method of integration by parts:

$$\int x \cos x \, dx.$$

Answer: _____

Check your answer in 322.

322

$$\int x \cos x \, dx = x \sin x + \cos x + c$$

If you want to see the derivation of this, continue here. Otherwise, go on to 323.

Let us make the following substitution and integrate by parts:

$u = x$, $dv = \cos x \, dx$. Thus $du = dx$, $v = \sin x$.

$\int x \cos x \, dx = \int u \, dv = uv - \int v \, du =$

$x \sin x - \int \sin x \, dx = x \sin x + \cos x + c.$

Go to 323.

Answers: (319) $(x - 1) \, e^x$

323

In integration problems it is often necessary to use a number of different integration procedures in a single problem.

Try the following: (b is a constant)

(a) $\int [\cos 5\theta + b]\, d\theta =$ _____

(b) $\int x \ln x^2\, dx =$ _____

Go to 324 for the answers.

324

The correct answers are

(a) $\int [\cos 5\theta + b]\, d\theta = \dfrac{1}{5} \sin 5\theta + b\theta + c$

(b) $\int x \ln x^2\, dx = \dfrac{1}{2} [x^2 (\ln x^2 - 1) + c]$

If you did both of these correctly you are doing fine — jump ahead to Section 3, frame 326. If you missed either problem, go to frame 325.

325

If you missed (a), you may have been confused by the change in notation from x to θ. Remember x is just a general symbol for a variable. All of the integration formulae could be written with θ, or z, or whatever you wish replacing the x. Now for (a) in detail:

$$\int [\cos 5\theta + b] \, d\theta = \int \cos 5\theta \, d\theta + \int b \, d\theta$$

$$= \frac{1}{5} \int \cos 5\theta \, d(5\theta) + \int b d\theta$$

$$= \frac{1}{5} \sin 5\theta + b\theta + c.$$

For problem (b), let $u = x^2$, $du = 2x \, dx$:

$$\int x \ln x^2 \, dx = \frac{1}{2} \int \ln u \, du = \frac{1}{2} [u \ln u - u + c].$$

(The last step uses formula 9, frame 307.) Therefore,

$$\int x \ln x^2 \, dx = \frac{1}{2} [x^2 \ln x^2 - x^2 + c].$$

You could also have solved this problem by integration by parts.

Go to Section 3, frame 326.

Section 3. THE AREA UNDER A CURVE

326

One important use of integration is in finding the area between some curve $y = f(x)$ and the x-axis, limited by two values of x, a and b. The shaded area in the figure is an example.

In this section you are going to learn how to compute that area.

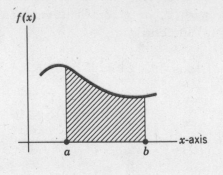

Go to 327.

327

Just to make sure you see what is meant, you should be able to calculate the area under the simplest of all possible curves — a straight line given by

$$f(x) = \text{constant}.$$

What is the area under the line $f(x) = 3$, between two points on the x-axis which have values, a and b?

$$A = \boxed{3ab \mid 3(a + b) \mid 3(a - b) \mid 3(b - a)}$$

If correct, go to 329.
Otherwise, go to 328.

328

The region indicated is a rectangle. The area of a rectangle is simply the product of its height and width. In this
case, the height is 3 and the width is the distance $b - a$. Thus:

$$A = 3(b - a).$$

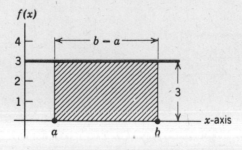

Go to 329.

Answers: (327) $3(b - a)$

329

Before continuing, we should point out that with this definition of area under a curve, area can be either positive or negative. To show this let's consider a few examples in which the
areas are rectangular for simplicity.

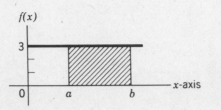

We will let A_{ab} represent the
area in the rectangle bounded
by $f(x) = 3$, and the x-axis,
with a base extending from $x =
a$ to $x = b$. Clearly, $A_{ab} = 3 \times
(b - a)$. If we now consider
A_{ba}, the same rectangle but
with the base extending from
$x = b$ to $x = a$, we have $A_{ba} = 3 \times (a - b) = -A_{ab}$. In this drawing A_{ab} is positive, and A_{ba} is negative.

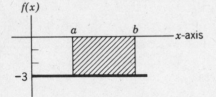

The height of the rectangle
can also be negative, as in
the drawing on the left. Here
A_{ab} is negative, and A_{ba} is
positive.

Even if the area is not
rectangular, the same
ideas hold.   In the
drawing to the right,

$A_{cd} > 0, A_{ef} < 0,$

$A_{dc} < 0, A_{fe} > 0.$

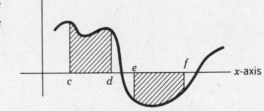

Although we will not generally denote the base points with
subscripts as we have done here in A_{ab}, they will always be
clear from the problem.

Go to 330.

330

We were able to find the area under a straight line since the figure formed is a simple rectangle. Now let's seek a general method for obtaining the area under any curve.

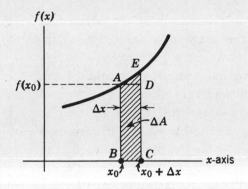

To begin, we will find the approximate area under any curve, $y = f(x)$, between two points on the x-axis which are separated by some *small* distance, Δx. If Δx is small, the area lies in a narrow strip, and is also small. We will denote it by ΔA. Can you find an approximate expression for ΔA?

$$\Delta A \approx \underline{\hspace{6cm}}$$

($\approx$ means "is approximately equal to")

This represents a major step in the development of integral calculus, so don't be disappointed if you need help.

For the correct answer, go to 331.

331

The answer we want is

$$\Delta A \approx f(x_0)\,\Delta x$$

If you wrote this, go on to 332. Otherwise, read below.

Let's take a close look at the area. As you can see, the area is a long thin strip. Unfortunately it is not a rectangle — but is nearly so. Most of the area is that of the rectangle $ABCD$ and this area is the product of the length $f(x_0)$ and its width Δx, that is $f(x_0)\,\Delta x$. The desired area ΔA differs from the area of the rectangle by the area of the figure ADE, which is almost a triangle except that the side AE is not straight. When the value of Δx becomes smaller and smaller, the area of the figure ADE becomes smaller at

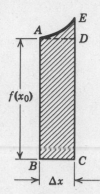

an even more rapid rate because both its base AD *and* its height DE become smaller in contrast to the rectangle $ABCD$ for which the length $f(x_0)$ stays fixed and only the width, $BC = \Delta x$, decreases. (Perhaps this argument has a familiar ring to it. What we are implying is that the approximation approaches an equality in the *limit* where $\Delta x \to 0$. More precisely, $\displaystyle\lim_{\Delta x \to 0}\ \frac{\Delta A}{f(x_0)\,\Delta x} = 1$.)

For a sufficiently small value of Δx, then, we can say

$$\Delta A \approx f(x_0)\,\Delta x.$$

Note that with similar accuracy we could have said

$$\Delta A \approx f(x_0 + \Delta x)\,\Delta x$$

and with even greater accuracy

$$\Delta A \approx f\left(x_0 + \frac{\Delta x}{2}\right)\Delta x.$$

However, the first of these is the simplest and is sufficiently accurate if Δx is small enough.

Go to 332.

332

Now let's see how to find the area under the curve $f(x)$ lying between vertical lines crossing the x-axis at a and x. Obviously, the area will depend on the value of x as well as on a and $f(x)$ so we will write it for the present as $A(x)$. With some thought you may be able to find an expression for $\dfrac{dA(x)}{dx}$.

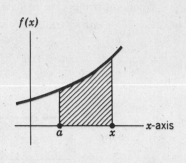

Try to figure this out (you will have to use the last frame) and then see frame 333 for the correct procedure.

$$\frac{dA(x)}{dx} = \underline{\hspace{3cm}}$$

Go to 333.

333

Here is how to work the problem formally.

$$\frac{dA(x)}{dx} = \lim_{\Delta x \to 0} \frac{A(x + \Delta x) - A(x)}{\Delta x}$$

But $A(x + \Delta x) - A(x)$ is the area ΔA of the strip shown. If Δx is small, we can use the result of frame 331. The area of the strip is approximately $\Delta A = f(x)\,\Delta x$.

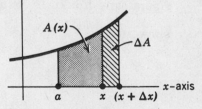

Now we can take the limit

$$\lim_{\Delta x \to 0} \frac{A(x + \Delta x) - A(x)}{\Delta x} = \lim_{\Delta x \to 0} \frac{\Delta A}{\Delta x} = \lim_{\Delta x \to 0} \frac{f(x)\,\Delta x}{\Delta x}$$

$$= \lim_{\Delta x \to 0} f(x) = f(x)$$

We have our result! $\boxed{\dfrac{dA(x)}{dx} = f(x)}$

Go to 334.

334

    To illustrate the result we have just obtained, let's consider a case we know exactly — the area $A(x)$ under a straight line.

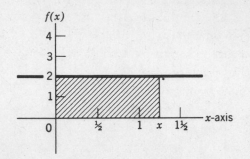

    Here is a graph of $f(x) = 2$. The area below the straight line $f(x) = 2$ and the x-axis and bounded by vertical lines at 0 and x, is obviously $2x$, as we can see from the diagram. Hence we have $A(x) = 2x$, from which it follows by differentiation that $\dfrac{dA}{dx} = 2$.

But $f(x) = 2$. In this case $\dfrac{dA}{dx} = f(x)$, as we showed in the last frame to be true in general.

    Now try this problem: Here is a graph of $f(x) = \dfrac{1}{2}\, x$.   Find the area between it and the x-axis bounded by vertical lines at 1 and x, as in the figure, and from this result show that $\dfrac{dA}{dx} = f(x)$.

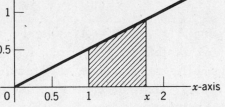

To check your result.
go to 335.

335

The area $A(x)$ is that of a right triangle of base $b = x$ and height $h = f(x) = \frac{1}{2} x$, minus the area of the smaller triangle which has base $= 1$ and height $= \frac{1}{2}$.

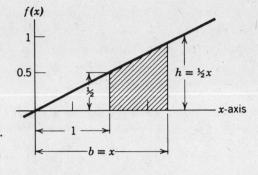

Hence, $A = \frac{1}{2} bh - \frac{1}{2} (\frac{1}{2} \times 1) = \frac{1}{4} x^2 - \frac{1}{4}$

$$\frac{dA}{dx} = \frac{d}{dx} (\frac{1}{4} x^2 - \frac{1}{4}) = \frac{1}{2} x = f(x).$$

Again we see that the result

$$\frac{dA(x)}{dx} = f(x)$$

holds true.

In fact, this result is quite general, and holds for the area under any shaped curve, as indeed we proved in frame 333. These two examples should make the result seem more plausible.

Go to 336.

336

We still have not solved our problem. We set out to find the area $A(x)$ and we have only found its derivative. Now you may see why the idea of an integral is so important.

Since $\frac{dA}{dx} = f(x)$, then

$$A(x) = \int f(x) \, dx$$

This expression is correct but is not yet very useful. Can you see why?

Go to 337.

337

The expression is not very useful because the indefinite integral involves an arbitrary constant whereas there is nothing indefinite or arbitrary about the area. Let's see how to get around this problem. Let the desired area $A(x)$ extend from a to x as in the drawing.

Suppose we have found a particular integral of $f(x)$, say

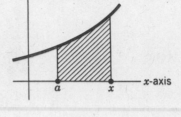

$$F(x) = \int f(x)\, dx$$

We also know

$$A(x) = \int f(x)\, dx.$$

However, as we discussed in frame 296, integrals of the same function can differ only by a constant. Since $A(x)$ and $F(x)$ are two integrals of $f(x)$, it follows that

$$A(x) = F(x) + c$$

The next problem is to find the correct value of the constant c. Can you?

$c =$ _____

If you need a hint, go to 338.
*If you would like to check your
solution, go to 339.*

338

Hint — Consider $A(a)$, the value of $A(x)$ when $x = a$.

Now — what is c ?

$c =$ _____

See 339 for the answer.

339

Consider the shaded area under $f(x)$ with edges at a and x in the limiting case when $x = a$, that is, when the two edges of the strip coincide. Clearly in this case the area is 0 since the strip has no width. Hence,

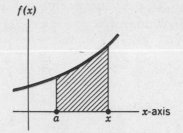

$A(a) = 0$. However, from frame 337

$A(x) = F(x) + c$. Consequently

$\qquad A(a) = F(a) + c = 0$

so $\qquad c = -F(a)$.

Go to 340.

340

At last we have our desired result.

$\qquad A(x) = F(x) + c$, and $c = -F(a)$, so

$$\boxed{A(x) = F(x) - F(a)}$$

In this equation recall that

$A(x) =$ area under $f(x)$ between points a and x,

$\quad a =$ some given value of x,

$F(x) =$ any indefinite integral of x

$\qquad = \int f(x)\, dx$,

$F(a) = F(x)$ evaluated at $x = a$.

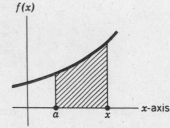

Go to 341.

341

To see how all this works, we will find the area under the curve $y = x^2$ between $x = 0$ and some value of x.

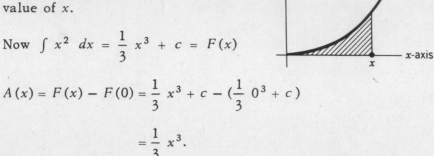

Now $\int x^2\ dx = \dfrac{1}{3} x^3 + c = F(x)$

$$A(x) = F(x) - F(0) = \frac{1}{3} x^3 + c - (\frac{1}{3}\ 0^3 + c)$$

$$= \frac{1}{3} x^3.$$

Note that the undetermined constant c drops out, as indeed it must. This occurs whenever we evaluate an expression such as $F(x) - F(a)$, so we can simply omit the c. We'll do this in the next few frames.

Go to 342.

342

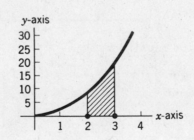

Can you find the area under the curve $y = 2x^2$, between the points $x = 2$ and $x = 3$?

$$A = \boxed{13 \mid \frac{1}{3} \mid \frac{38}{3} \mid 18}$$

If right, go to 344.
Otherwise, go to 343.

343

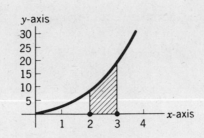

Here is how to solve the problem:

$$A = F(3) - F(2), \quad F(x) = \int 2x^2 \, dx = \frac{2}{3} x^3$$

$$A = \frac{2}{3} \times 27 - \frac{2}{3} \times 8 = 18 - \frac{16}{3} = \frac{38}{3}.$$

Go to 344.

344

Before we go on, let's introduce a little labor saving notation.

Frequently we have to find the difference of an expression evaluated at two points, as $F(b) - F(a)$. This is often denoted by

$$F(b) - F(a) = F(x) \Big|_a^b$$

For instance, $x^2 \Big|_a^b = b^2 - a^2$.

As another example, in the last problem, we had to evaluate

$$\frac{2}{3} x^3 \Big|_2^3 = \frac{2}{3} (3^3) - \frac{2}{3} (2^3) = \frac{2}{3} (27 - 8) = \frac{38}{3}$$

Go to 345.

Answers: (342) $\dfrac{38}{3}$

345

Let's do one more practice problem:

The graph shows a plot of $y = x^3 + 2$.

Find the area between the curve and the x-axis from $x = -1$ to $x = +2$.

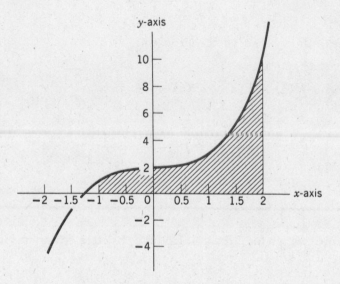

Answer: $\boxed{5 \mid 1/4 \mid 4 \mid 17/4 \mid 39/4 \mid \text{none of these}}$

If right, go to Section 4,
frame 347.
Otherwise, go to 346.

346

Here is how to do the problem:

$$A = F(2) - F(-1) = F(x) \Big|_{-1}^{2}$$

$$F = \int y \, dx = \int (x^3 + 2)\, dx = \frac{1}{4} x^4 + 2x$$

$$A = \left(\frac{1}{4} x^4 + 2x\right)\Big|_{-1}^{2} = \left(\frac{16}{4} + 4\right) - \left(\frac{1}{4} - 2\right) = \frac{39}{4}$$

Go to 347.

Section 4.   DEFINITE INTEGRALS

347

In this section we are going to find another way to compute the area under a curve. Our new result will be equivalent to that of the last section, but it will give us a new point of view.

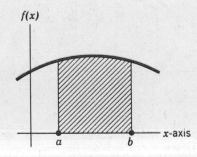

Let's briefly summarize the last section. If A is the area under the curve of $f(x)$ between $x = a$ and some value x, then we showed (frame 333) that $\dfrac{dA}{dx} = f(x)$. From this we went on to show (frame 340) that if $F(x)$ is an indefinite integral of $f(x)$, i.e., $\dfrac{dF}{dx} = f$, then the area under $f(x)$ between the two values of x, a and b, is given by

$$A = F(b) - F(a).$$

Now for a new approach!

Go to 348.

Answer: (345) $\dfrac{39}{4}$

348

    Let's evaluate the area under a curve in the following manner:

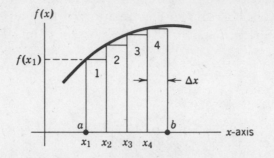

    First we divide the area into a number of strips of equal widths by drawing lines parallel to the axis of $f(x)$. The figure shows 4 such strips drawn. The strips have irregular tops, but we can make them rectangular by drawing a horizontal line at the top of each strip as shown. Suppose we label the strips by 1, 2, 3, 4. The width of each strip is

$$\Delta x = \frac{b - a}{4}.$$

The height of the first strip is $f(x_1)$, where x_1 is the value of x at the beginning of the first strip. Similarly, the height of strip 2 is $f(x_2)$, where $x_2 = x_1 + \Delta x$. The third and fourth strips have heights $f(x_3)$ and $f(x_4)$ respectively, where $x_3 = x_1 + 2\Delta x$, and $x_4 = x_1 + 3\Delta x$.

Go to 349.

349

    You should be able to write an approximate expression for the area of any of the strips. If you need help, review frame 331. Below write the approximate expression for the area of strip number 3, ΔA_3,

$\Delta A_3 \approx$ _____

For the correct answer, go to 350.

350

The approximate area of strip number 3 is $\Delta A_3 \approx f(x_3)\,\Delta x$.

If you want to see a discussion of this, refer again to frame 331.

Can you write an approximate expression for A, the total area of all 4 strips?

$$A \approx \underline{\hspace{5cm}}$$

Try this, and then see 351
for the correct answer.

351

An approximate expression for the total area is simply the sum of the areas of all the strips. In symbols, since $A = \Delta A_1 + \Delta A_2 + \Delta A_3 + \Delta A_4$, we have

$$A \approx f(x_1)\,\Delta x + f(x_2)\,\Delta x + f(x_3)\,\Delta x + f(x_4)\,\Delta x.$$

We could also write this

$$A \approx \sum_{i=1}^{4} f(x_i)\,\Delta x.$$

Σ is the Greek letter *sigma* which corresponds to the English letter S and stands here for the sum. The symbol $\sum_{i=1}^{n} g(x_i)$ means

$$g(x_1) + g(x_2) + g(x_3) + \ldots + g(x_n).$$

Go to 352.

352

Suppose we divide the area into more strips each of which is narrower, as shown in the drawings. Evidently our approximation gets better and better.

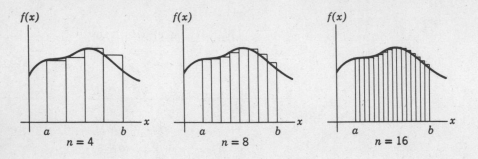

If we divide the area into n strips, then $A \approx \sum\limits_{i=1}^{n} f(x_i)\, \Delta x$, where $n = \dfrac{b-a}{\Delta x}$. Now, if we take the limit where $\Delta x \to 0$, the approximation becomes an equality. Thus,

$$A = \lim_{\Delta x \to 0} \sum_{i=1}^{n} f(x_i)\, \Delta x.$$

Such a limit is so important that it is given a special name and symbol. It is called the *definite integral* and is written $\int_a^b f(x)\, dx$. This symbol looks similar to the indefinite integral, $\int f(x)\, dx$, and, as we shall see in the next frame, it is related. However, it is important to remember that the definite integral is defined by the limit described above. So, by definition

$$\boxed{\int_a^b f(x)\, dx = \lim_{\Delta x \to 0} \sum_{i=1}^{n} f(x_i)\, \Delta x}$$

(Incidentally, the symbol $\int$ evolved from the letter S and like sigma it was chosen to stand for *sum*.)

Go to 353.

353

    With this definition for the definite integral, the discussion in the last frame shows that the area A under the curve is equal to the *definite* integral.

$$A = \int_a^b f(x)\ dx$$

But we saw earlier that the area can also be evaluated in terms of the *indefinite* integral.

$$F(x) = \int f(x)\ dx$$

by

$$A = F(b) - F(a).$$

Therefore we have the general relation

$$\int_a^b f(x)\ dx = F(b) - F(a) = \left\{ \int f(x)\ dx \right\}\Bigg|_a^b$$

Thus the *definite* integral can be expressed in terms of an *indefinite* integral evaluated at the limits. This remarkable result is often called the Fundamental Theorem of Integral Calculus.

Go to 354.

354

    To help remember the definition of definite integral try writing it yourself. Write an expression defining the definite integral of $f(x)$ between limits a and b.

To check your answer,
go to 355.

355

The correct answer is

$$\int_a^b f(x)\, dx = \lim_{\Delta x \to 0} \sum_{i=1}^{n} f(x_i)\, \Delta x, \text{ where } n = \frac{b-a}{\Delta x}.$$

Congratulations if you wrote this, or an equivalent expression.

If you wrote

$$\int_a^b f(x)\, dx = F(b) - F(a), \text{ where } F(x) = \int f(x)\, dx,$$

your statement is true, but it is not the *definition* of a definite integral. The result is true because both sides represent the same thing—the area under the curve of $f(x)$ between $x = a$ and $x = b$. It is an important result, since without it we would have no way of evaluating the definite integral, but it is not true by definition.

If this reasoning is clear to you, go right on to 356.

Otherwise, review the material in this chapter, and then, to see a further discussion of definite and indefinite integrals go on to 356.

356

Perhaps the definite integral seems an unnecessary complication to you. After all, the only thing we accomplished with it was to write the area under a curve a second way. To actually compute the area we were led back to the indefinite integral. However, we could have found the area directly from the indefinite integral in the first place. The importance of the definite integral arises from its definition as the limit of a sum. The process of dividing a system into little bits and then adding them all together is applicable to many problems. This naturally leads to definite integrals which we can evaluate in terms of indefinite integrals by using the Fundamental Theorem in frame 353.

Go to 357.

357

Can you prove that

$$\int_a^b f(x)\, dx = -\int_b^a f(x)\, dx\,?$$

After you have tried to prove
this result, go to 358.

358

The proof that $\int_a^b f(x)\, dx = -\int_b^a f(x)\, dx$ is simple.

$$\int_a^b f(x)\, dx = F(b) - F(a),\ \text{where}\ F(x) = \int f(x)\, dx$$

but

$$\int_b^a f(x)\, dx = F(a) - F(b) = -\,[F(b) - F(a)]$$

$$= -\int_a^b f(x)\, dx.$$

It is evident from frame 329 that reversing the points a and b reverses the sign of the area.

The points a and b are called the limits of the integral (nothing to do with $\lim\limits_{x \to a} f(x)$; here limit simply means the boundary). The process of evaluating

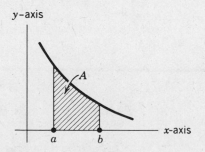

$$\int_a^b f(x)\, dx$$

is often spoken of as "integrating $f(x)$ from a to b" and the expression is called the "integral of $f(x)$ from a to b".

Go to 359.

359

Which of the following expressions correctly gives $\int_0^{2\pi} \sin \theta \, d\theta$?

$$\boxed{1 \mid 0 \mid 2\pi \mid -2 \mid -2\pi \mid \text{none of these}}$$

Go to 360.

360

$$\int_0^{2\pi} \sin \theta \, d\theta = -\cos \theta \,\Big|_0^{2\pi} = -[1 - 1] = 0$$

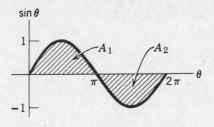

It is easy to see why this result is true by inspecting the figure. The integral yields the total area under the curve, from 0 to 2π, which is the sum of A_1 and A_2. But A_2 is negative, since $\sin \theta$ is negative in that region. By symmetry, the two areas just add to 0. However, you should be able to find A_1 or A_2 separately. Try this problem:

$$A_1 = \int_0^{\pi} \sin \theta \, d\theta = \boxed{1 \mid 2 \mid -1 \mid -2 \mid \pi \mid 0}$$

If right, go to 362.
Otherwise, go to 361.

361

$$A_1 = \int \sin \theta \, d\theta = -\cos \theta \,\Big|_0^{\pi} = -[-1 - (+1)] = 2.$$

If you forgot the integral, you can find it in the table on page 285. In evaluating $\cos \theta$ at the limits, we need to know that $\cos (\pi) = -1$, $\cos (0) = 1$.

Go to 362.

362

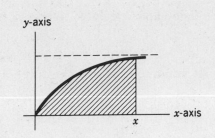

Here is a graph of the function $y = 1 - e^{-x}$.

Can you find the shaded area under the curve between the origin and x?

Answer: $\boxed{e^{-x} \mid 1 - e^{-x} \mid x + e^{-x} \mid x + e^{-x} - 1}$

> *Go to 364 if you did this correctly.*
> *See 363 for the solution, or if you want*
> *to see a discussion of the meaning of the area.*

363

Here is the solution to 362.

$$A = \int_0^x y\,dx = \int_0^x (1 - e^{-x})\,dx = \int_0^x dx - \int_0^x e^{-x}\,dx$$

$$= x - (-e^{-x}) \Big|_0^x = x + e^{-x} \Big|_0^x = x + e^{-x} - 1,$$

The area found is bounded by a vertical line through x. Our result gives A as a variable that depends on x. If we choose a specific value for x, we can substitute it into the above formula for A and obtain a specific value for A. We have evaluated a definite integral in which one of the boundary points has been left as a variable.

> *Go to 364.*

Answers: (359) 0;   (360) 2

364

Let's evaluate one more definite integral before going on. Find

$$\int_0^1 \frac{dx}{\sqrt{1 - x^2}}$$ (If you need to, use the integral tables, p 285.)

Answer: $\left[\,0 \mid 1 \mid \infty \mid \pi \mid \dfrac{\pi}{2} \mid \text{none of these}\,\right]$

If you got the right answer, go to section 5, frame 366.
If you got the wrong answer, or no answer at all, go to 365.

365

From the integral table, p 285, we see that

$$\int \frac{dx}{\sqrt{1 - x^2}} = \arcsin x + c.$$ Therefore,

$$\int_0^1 \frac{dx}{\sqrt{1 - x^2}} = \arcsin x \,\Big|_0^1 = \arcsin 1 - \arcsin 0.$$

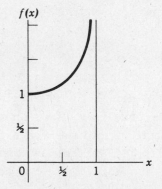

But arcsin $(1) = \pi/2$, since sin $(\pi/2) = 1$. Similarly, arcsin $(0) = 0$. Thus, the integral has the value $\pi/2 - 0 = \pi/2$.

A graph of $f(x) = \dfrac{1}{\sqrt{1 - x^2}}$ is shown at the left. Although the function is discontinuous at $x = 1$, the area under the curve is perfectly well defined.

Go to section 5, frame 366.

Answer: (362) $x + e^{-x} - 1$

Section 5.   SOME APPLICATIONS OF INTEGRATION

366

In this section we are going to apply integration to a few simple problems.

In Chapter II we learned how to find the velocity of a particle if we knew its position in terms of time. Now we can reverse the procedure and find the position from the velocity. For instance, we are in an automobile driving along a straight road through thick fog. To make matters worse, our mileage indicator is broken. Instead of watching the road all the time, let's keep an eye on the speedometer. We have a good watch along, and we make a continuous record of the speed starting from the time when we were at rest. The problem is to find how far we have gone. (This is a dangerous method for navigating a car, but it is actually used for navigating submarines and spacecraft.) More specifically, given $v(t)$, how do we find $S(t)$, the distance travelled since time t_0 when we were at rest? Try to work out a method.

$S(t) = $ _____

To check your result,
go to 367.

Answer: (364) $\pi/2$.

367

Since

$$v = \frac{dS}{dt}$$

we must have $dS = v\, dt$ (as was shown in 266).

Now let us integrate both sides from the initial point ($t = t_0$, $S = 0$) to the final point (t, S).

We have

$$\int_{S=0}^{S} dS = \int_{t_0}^{t} v\, dt, \text{ so}$$

$$S = \int_{t_0}^{t} v\, dt.$$

*If you did not get this result, or would
like to see more explanation,
go to 368.
Otherwise, go to 369.*

368

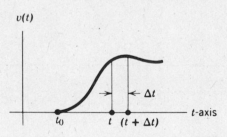

Another way to understand this problem is to look at it graphically. Here is a plot of $v(t)$ as a function of t. In time Δt the distance travelled is $\Delta S = v\, \Delta t$. The total distance travelled is thus equal to the area under the curve between the initial time and the time of interest and this is $\int_{t_0}^{t} v(t)\, dt$.

Go to 369.

369

Suppose an object moves with a velocity which continually decreases in the following way.

$$v(t) = v_0\, e^{-bt}.$$

v_0 and b are constants.

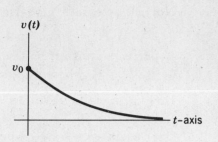

At $t = 0$ the object is at the origin; $S = 0$. Which of the following is the distance the object will have moved after an infinite time (or, if you prefer, after a very long time)?

$$\left[\, 0 \mid v_0 \mid v_0 e^{-1} \mid \frac{v_0}{b} \mid \infty \,\right]$$

If correct, go to 371.
Otherwise, go to 370.

370

Here is the solution to the problem of frame 369.

$$S(t) - S(0) = \int_0^t v\, dt = \int_0^t v_0\, e^{-bt}\, dt$$

$$S(t) - 0 = -\frac{v_0}{b}\, e^{-bt}\,\Big|_0^t = -\frac{v_0}{b}(e^{-bt} - 1)$$

We are interested in $\lim_{t \to \infty} S(t)$, but since $e^{-bt} \to 0$ as $t \to \infty$, we have

$$\lim_{t \to \infty} S(t) = -\frac{v_0}{b}(0 - 1) = \frac{v_0}{b}$$

Although the object never comes completely to rest, its velocity gets so small that the total distance travelled is finite.

Go to 371.

371

Not all integrals give finite results. For example, try this problem.

A particle starts from the origin at $t = 0$ with a velocity $v(t) = v_0/(b + t)$, where v_0 and b are constants.

How far does it travel as $t \to \infty$?

$$\left[v_0 \ln \frac{1}{b} \mid \frac{v_0}{b} \mid \frac{v_0}{b^2} \mid \text{ none of these} \right]$$

Go to 372.

372

It is easy to see that problem 371 leads to an infinite integral.

$$S(t) - 0 = \int_{t=0}^{t} v_0 \frac{dt}{b + t} = v_0 \ln (b + t) \Big|_0^t$$

$$= v_0 \left[\ln (b + t) - \ln b \right]$$

$$= v_0 \ln \left(1 + \frac{t}{b}\right)$$

Since $\ln \left(1 + \frac{t}{b}\right) \to \infty$ as $t \to \infty$, we see that $S(t) \to \infty$ as $t \to \infty$.

In this case, the particle is always moving fast enough so that its motion is unlimited. Or, alternatively, the area under the curve $v(t) = v_0/(b + t)$ increases without limit as $t \to \infty$.

Go to 373.

Answer: (369) $\dfrac{v_0}{b}$

373

In the next few frames we are going to apply integration to find the volume of a right circular cone.

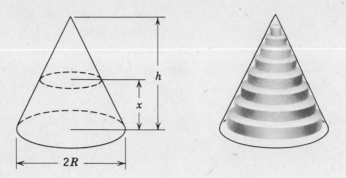

The height of the cone is h, and the radius of the base is R. We will let x represent distance vertically from the base.

Our method of attack is similar to that used in frame 348 to find the area under a curve. We will slice the body into a number of discs whose volume is approximately that of the cone in the figure (the cone has been approximated by 8 circular discs). Then we have

$$V = \sum_{i=1}^{8} \Delta V_i$$

where ΔV_i is the volume of one of the discs. In the limit where the height of each disc (and hence the volume) goes to 0, we have

$$V = \int dV.$$

In order to evaluate this, we have to have an expression for dV. To find this, go to frame 374.

Answer: (371) none of these

374

Because we are going to take the limit where $\Delta V \to 0$, we will represent the volume element by dV from the start.

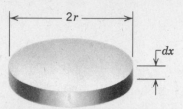

Here is a picture of a section of the cone, which for our purposes is represented by a disc. The radius of the disc is r and its height is dx. Try to find an expression for dV in terms of x. (You will have to find r in terms of x).

$dV =$ _____

To check your result, or to see how to obtain the result, go to 375.

375

$$dV = \pi R^2 \left(1 - \frac{x}{h}\right)^2 dx.$$

If you got this answer, go on to 376.

If you want to see how to derive it, read on.

The volume of this disc is the product of the area and height. Thus, $dV = \pi r^2 \, dx$. Our remaining task is to express r in terms of x.

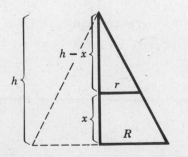

The diagram shows a cross section of the cone. Since r and R are corresponding edges of similar triangles, it should be clear that $\frac{r}{R} = \frac{h - x}{h}$, or $r = R\left(1 - \frac{x}{h}\right)$. Thus,

$$dV = \pi R^2 \left(1 - \frac{x}{h}\right)^2 dx.$$

Go to 376.

376

We now have an integral for V.

$$V = \int_0^h dV = \int_0^h \pi R^2 \left(1 - \frac{x}{h}\right)^2 dx.$$

Try to evaluate this.

$V =$ _____

To check your answer,
go to 377.

377

You should have obtained the result

$$V = \frac{1}{3} \pi R^2 h.$$

Congratulations, if you did. Go on to 378. Otherwise, read below

$$V = \int_0^h \pi R^2 \left(1 - \frac{x}{h}\right)^2 dx = \pi R^2 \int_0^h \left(1 - \frac{2x}{h} + \frac{x^2}{h^2}\right) dx.$$

$$= \pi R^2 \left[x - \frac{x^2}{h} + \frac{1}{3}\frac{x^3}{h^2}\right] \Big|_0^h = \pi R^2 \left(h - h + \frac{1}{3} h\right)$$

$$= \frac{1}{3} \pi R^2 h.$$

Go to 378.

378

Here is one more problem. Let's find the volume of a sphere.

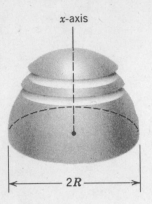

x-axis

2*R*

It will simplify matters if we find the volume of a hemisphere, V', which is just half the required volume, V. Thus, $V' = V/2$.

Can you write an integral which will give the volume of the hemisphere? (The slice of the hemisphere shown in the drawing may help you in this.)

$$V' = \underline{\hspace{5cm}}$$

Go to 379 to check your formula.

379

You should have written

$$V' = \int_0^R \pi \, (R^2 - x^2) \, dx.$$

If you wrote this, go ahead to frame 380.
Otherwise, continue here.

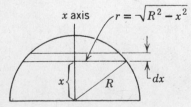

x axis $r = \sqrt{R^2 - x^2}$

x   *R*   *dx*

Here is a vertical section through the hemisphere. The volume of the disc between x and $x + dx$ is $\pi \, r^2 \, dx$. But, as can be seen from the triangle indicated, $x^2 + r^2 = R^2$ so
$$r^2 = R^2 - x^2.$$

Hence, $dV' = \pi \, (R^2 - x^2) \, dx$ and $V' = \int_0^R \pi \, (R^2 - x^2) \, dx$.

Go to 380.

380

Now go ahead and evaluate the integral

$$V' = \int_0^R \pi\,(R^2 - x^2)\,dx.$$

$$V' = $$

_____.

To see the correct answer,
go to 381.

381

$$V' = \int_0^R \pi\,(R^2 - x^2)\,dx = \pi\,[R^2 x - \frac{1}{3}\,x^3]\,\Big|_0^R$$

$$= \pi\,(R^3 - \frac{1}{3}\,R^3) = \frac{2}{3}\,\pi\,R^3.$$

Since $V = 2\,V'$, $V = \frac{4}{3}\,\pi\,R^3$

Go to Section 6,
frame 382.

Section 6.   MULTIPLE INTEGRALS

382

The subject of this section, multiple integrals, is interesting, but it is also a little complicated and, depending on your particular interests, may not be essential for your later work. Therefore, if you feel you have had about as much calculus as you want right now you should skip on to the conclusion of this chapter, frame 400. Otherwise, continue here.

So far we have considered simple integrals such as $\int_a^b f(x)\,dx$. Now we are going to study a more general kind of integral. To start, let's find new ways to look at a familiar problem.

Suppose we want to find the area inside a closed curve. The curve may be defined by more complicated expressions than we have had so far. For example, the curve given by

$$x^2 + y^2 - r^2 = 0$$

is a circle of radius r centered on the origin. The curve shown in the figure represents some other equation which relates x and y. Our problem is to find the enclosed area, A.

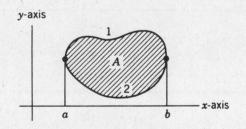

We could find the area under curve (1) between a and b, and then subtract the area under curve (2), but there are other ways to do it.

To find out how,
go to 383.

383

Let's start by dividing the area into strips, just as we did in frame 348 where we found the area between a curve and the x-axis. The basic difference is that here the co-ordinates of both ends of the strip depend on the value of x at the strip. Thus if the width of each strip is Δx, the area of the shaded strip is $\Delta x \times (y_2 - y_1)$.

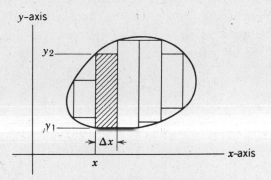

If we let $y_2(x)$ represent the value of y at the top of a strip located at x and $y_1(x)$ represent the value of y at the bottom, then using the same arguments as in frame 352, we have

$$A = \lim_{\Delta x \to 0} \sum_{i=1}^{n} [y_2(x_i) - y_1(x_i)]\, \Delta x = \int_a^b [y_2(x) - y_1(x)]\, dx$$

Let us now consider still a different way of expressing A.

To do this,
go to 384.

384

Here is a picture of one of the strips making up the area of the figure in 383. We can subdivide the strips into smaller areas composed of rectangles of height Δy with the same widths as the strip, Δx. Then the area of the strip is approximately the sum of the area of the rectangles, as shown. Area of strip $\approx \Sigma_y \Delta y \Delta x$. For simplicity we won't show the limits of the sum here. Instead, we will simply remember that the sum goes from the minimum value of y, y_1, to the maximum value, y_2. The subscript y below Σ is to remind us that we are summing all the $\Delta y's$, multiplied by a fixed Δx.

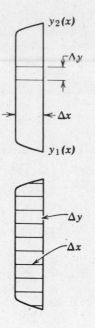

Our next step is to write an exact expression for the total area enclosed by the curve in terms of the limits of two sums, one over x and the other over y. Try to do this. For simplicity omit the limits on the sums.

$$A = \underline{\hspace{4cm}}$$

To check your answer, or to see a discussion, go to 385.

385

The correct expression is

$$A = \lim_{\Delta x \to 0} \sum_x \left[\lim_{\Delta y \to 0} \sum_y \Delta y \right] \Delta x$$

$$= \lim_{\Delta x \to 0} \lim_{\Delta y \to 0} \sum_x \sum_y \Delta y \Delta x.$$

Congratulations if you wrote either of the above expressions. They are equivalent and are both correct. Whether or not you got this result, read on.

The second formula yields a particularly simple interpretation for the meaning of the double sum.

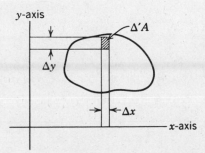

$\Delta y \Delta x$ is the area of the small rectangle shown, and we can equate this to $\Delta'A$. $\Delta'A$ is an increment of the area. The prime symbol (') is to remind us that $\Delta'A$ is the product of two small quantities, Δx and Δy. When we take the sum over x and y, we are in effect summing all the $\Delta'A$'s in the area. Thus, we could also write that formula as

$$A = \lim_{\Delta x \to 0} \lim_{\Delta y \to 0} \sum_x \sum_y \Delta'A.$$

However, in order to compute the area we will want to turn the sums into definite integrals, and to do that we will use the formula for A at the top of the page.

Go to 386.

386

Our result so far is

$$A = \lim_{\Delta x \to 0} \sum_x \{ \lim_{\Delta y \to 0} \sum_y \Delta y \} \, \Delta x.$$

The quantity in parentheses very much resembles a definite integral. In fact, using the result of frame 352, we have

$$\lim_{\Delta y \to 0} \sum_y \Delta y = \int_{y_1}^{y_2} dy$$

Here we explicitly insert the limits on y which have been omitted since frame 384. Remember that these limits, y_1 and y_2, depend on x. In case the use of dy seems strange to you, we will discuss it in the next frame. The definite integral above is easily evaluated:

$$\int_{y_1}^{y_2} dy = y_2 - y_1$$

However, we will leave it as an integral for the present, so that our new result is

$$A = \lim_{\Delta x \to 0} \sum_x \left[\int_{y_1}^{y_2} dy \right] \Delta x$$

Now you should be able to write A completely in terms of definite integrals. In doing this, let a represent the smallest value of x in the area, and let b represent the largest value.

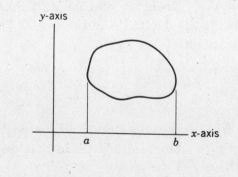

$$A = \underline{\hspace{5cm}}$$

To see the correct answer,
go to 387.

387

Our final result is

$$A = \int_a^b \left[\int_{y_1}^{y_2} dy \right] dx$$

When used in this context, both dx and dy are independent differentials. This was not the case in Section 11 of Chapter II. The reason is that although y depends on x on the curve we can regard x and y as independent variables when we integrate the area within the curve. The dependence of y on x enters the problem through the limits on the y integral. Both y_2 and y_1 depend on x.

Before going on, we should point out that the roles of x and y are interchangeable here. We could integrate x from one edge of the area to the other, and then integrate y over its permissible range.

For an example showing how this works in practice, go to 388.

388

Before trying a problem on your own, perhaps you would like to see one worked out in detail. Let's use this method to find the area of a circle given by $x^2 + y^2 - R^2 = 0$. The radius of the circle is R, and we know the answer beforehand: $A = \pi R^2$.

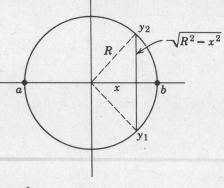

We have $A = \int_a^b \left[\int_{y_1}^{y_2} dy \right] dx$.

From the diagram, you should be able to see that $a = -R$, $b = R$, $y_1 = -\sqrt{R^2 - x^2}$, $y_2 = \sqrt{R^2 - x^2}$.

Thus $A = \int_{-R}^{+R} \left[\int_{-\sqrt{R^2 - x^2}}^{\sqrt{R^2 - x^2}} dy \right] dx$.

The y integral is

$$\int_{-\sqrt{R^2-x^2}}^{\sqrt{R^2-x^2}} dy = y \Big|_{-\sqrt{R^2 - x^2}}^{\sqrt{R^2 - x^2}} = 2\sqrt{R^2 - x^2}.$$

Substituting this into our formula for A, we have

$$A = 2 \int_{-R}^{+R} \sqrt{R^2 - x^2}\, dx.$$

This integral is not one we have so far come across. To see the rest of the solution go to 389.

389

Our task is to evaluate $\int_{-R}^{+R} \sqrt{R^2 - x^2}\,dx$.

This integral is not listed in the integrals of this book, but it is listed in more complete tables such as those referred to in Appendix B6. The result is

$$\int \sqrt{R^2 - x^2}\,dx = \frac{1}{2}\left[x\sqrt{R^2 - x^2} + R^2 \arcsin\frac{x}{R}\right].$$

You can check that this is true by differentiating the expression on the right. (Formula 19 of Table 1, p. 283, is helpful for this.)

Thus our result is

$$A = 2\int_{-R}^{+R} \sqrt{R^2 - x^2}\,dx = \left[x\sqrt{R^2 - x^2} + R^2 \arcsin\frac{x}{R}\right]\Big|_{-R}^{+R}$$

$$= R^2\left[\arcsin(1) - \arcsin(-1)\right]$$

Since $\arcsin(1) = \frac{\pi}{2}$ and $\arcsin(-1) = -\frac{\pi}{2}$, we have

$$A = R^2\left[\frac{\pi}{2} - \left(-\frac{\pi}{2}\right)\right] = \pi R^2.$$

Now for a problem you can try.

Go to 390.

390

Let's find the area in the triangle shown using the method developed in this section.

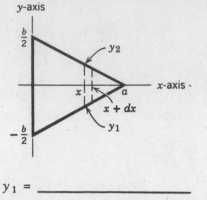

$$A = \int_{x_{min}}^{x_{max}} \int_{y_1}^{y_2} dy \, dx$$

The only real problem is to find the limits of integrals. (Can you?)

$y_2 =$ _____ $y_1 =$ _____

$x_{max} =$ _____ $x_{min} =$ _____

For the correct answer, see 391.

391

From the diagram you should be able to see

$$y_2 = \frac{b}{2} \left(1 - \frac{x}{a}\right)$$

$$y_1 = -\frac{b}{2} \left(1 - \frac{x}{a}\right).$$

(If you are in doubt about these, review frame 375).

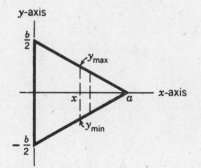

$$x_{max} = a \qquad\qquad x_{min} = 0$$

So $A = \int_0^a \left[\int_{-\frac{b}{2}(1-\frac{x}{a})}^{+\frac{b}{2}(1-\frac{x}{a})} dy \right] dx.$

Do the y integral next.

$$\int_{-\frac{b}{2}(1-\frac{x}{a})}^{\frac{b}{2}(1-\frac{x}{a})} dy = \rule{5cm}{0.4pt}$$

To check your result, go to 392.

392

The integral is simple, since

$$\int_{y_1}^{y_2} dy = y \Big|_{y_1}^{y_2} = y_2 - y_1$$

$$\int_{-\frac{b}{2}(1-\frac{x}{a})}^{+\frac{b}{2}(1-\frac{x}{a})} dy = \frac{b}{2}(1 - \frac{x}{a}) - [-\frac{b}{2}(1 - \frac{x}{a})]$$

$$= b(1 - \frac{x}{a}).$$

Now — complete the problem by doing the x integral.

$$A = \underline{\hspace{6cm}}$$

Go to 393.

393

$$A = \int_0^a b(1 - \frac{x}{a})\, dx = b(x - \frac{1}{2}\frac{x^2}{a}) \Big|_0^{+a}$$

$$= b(a - \frac{1}{2}\frac{a^2}{a}) = \frac{1}{2} ab.$$

This leads to a familiar result — namely that the area of the triangle $= \frac{1}{2}$ base $\times$ height.

Now — to see a less familiar problem

go to 394.

394

Suppose our triangle is the base of an object made of material with varying thickness. The thickness, z, varies in the following manner:

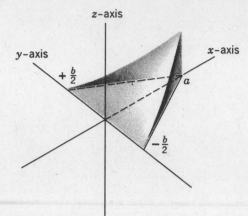

$z = C\,x^2$, where C is constant. (This is a peculiar piece of material. It has 0 thickness along the y-axis but the thickness increases rapidly as x increases.) The flat base of the object is on the x-y plane.

Our problem is to find the volume of the object.

Go to 395.

395

We proceed as before, but we now have three dimensions to sum over.

Let dV = volume of element with sides dx, dy, dz.

$$V = \iiint dV = \int \left\{ \int \left[\int dz \right] dy \right\} dx$$

(limits are omitted here for simplicity.)

Can you do the z integration? (Reread frame 392 if you are stuck on this).

$$\int_{z_1}^{z_2} dz = \underline{\hspace{4cm}}$$

Go to 396,
for the answer.

396

$$\int_{z_1}^{z_2} dz = z_{max} - z_{min} = C\,x^2 - 0 = C\,x^2$$

so
$$V = \int_{x_{min}}^{x_{max}} \left\{ \int_{y_1}^{y_2} C\,x^2\,dy \right\} dx$$

You should be able to carry through the problem. See if you
can, and then check the result in frame 397.

$$V =$$

<hr>

Go to 397.

397

The limits on x and y are just the same as in frame 390. So —

$$V = \int_0^a \left[\int_{-\frac{b}{2}(1-\frac{x}{a})}^{+\frac{b}{2}(1-\frac{x}{a})} C\,x^2\,dy \right] dx$$

$$= \int_0^a C\,x^2 \left[y \left|_{-\frac{b}{2}(1-\frac{x}{a})}^{+\frac{b}{2}(1-\frac{x}{a})} \right. \right] dx$$

$$= \int_0^a C\,x^2\,b\left(1 - \frac{x}{a}\right) dx = bC\left(\frac{1}{3}x^3 - \frac{1}{4}\frac{x^4}{a}\right)\Big|_0^a$$

$$= bC\left(\frac{1}{3}a^3 - \frac{1}{4}a^3\right) = \frac{bC\,a^3}{12}.$$

Go to 398.

398

You may feel that multiple integration is a rather complicated way to do something basically simple. In our expression for area,

$$A = \int [\int dy] \, dx$$

the y integration always led to $y_2 - y_1$, so that we obtained

$$A = \int (y_2 - y_1) \, dx,$$

which was just our starting point back in frame 384. If this were our only use for multiple integrals you would be quite correct. However, the use of multiple integrals is much more extensive than simply for evaluating areas. For instance, we can evaluate integrals in the form

$$G = \int_{x_{min}}^{x_{max}} \left[\int_{y_1}^{y_2} g(x,y) \, dy \right] dx$$

where $g(x,y)$ is a variable that depends on both x and y, and the limits y_1 and y_2 depend on x. We had an example of this in frame 397 where $g(x,y) = C x^2$. The procedure is straightforward: the limits on each of the integrals are found, and the y integration is performed with x treated like a constant in $g(x,y)$. Next, the x integration is done. The procedure is easily extended to integrals involving even more variables.

If you would like to see another example of such an integration, go to 399.

Otherwise, go to Section 7, frame 400.

399

In this frame we will evaluate

$$G = \iint (x + y)\, dy\, dx,$$

where the area is that enclosed in the semicircle shown. The y integral gives:

$$\int_{-\sqrt{R^2-x^2}}^{\sqrt{R^2-x^2}} (x + y)\, dy =$$

$$\left. (xy + \frac{1}{2} y^2) \right|_{-\sqrt{R^2-x^2}}^{\sqrt{R^2-x^2}} =$$

$$2x\sqrt{R^2-x^2}.$$

We then have

$$G = \int_0^R 2x\sqrt{R^2-x^2}\, dx.$$

We can evaluate this integral by substituting the new variable $u = x^2$. Then $du = 2x\, dx$, and we have

$$G = \int_0^{R^2} (R^2 - u)^{1/2}\, du.$$

(Note that we have substituted $u = x^2$ in the limits. Thus, at the upper limit, $x = R$ and $u = R^2$. Whenever we change a variable we must make the same change in the limits.)

If we make a new substitution, $s = R^2 - u$, and $ds = -du$, we obtain

$$G = -\int_{R^2}^0 s^{1/2}\, ds = +\int_0^{R^2} s^{1/2}\, ds = \left. \frac{2}{3} s^{3/2} \right|_0^{R^2}$$

$$= \frac{2}{3}(R^2)^{3/2} = \frac{2}{3} R^3.$$

Go to Section 7, frame 400.

Section 7.   CONCLUSION

400

At this point you should understand the principles of integration and be able to do some integrals. With practice your repertoire will increase. Don't be afraid to use integral tables — everyone does. You can find quite large tables in

The Handbook of Chemistry and Physics, Chemical Rubber Publishing Co., Cleveland, Ohio.

H. Dwight, *Tables of Integrals and Other Mathematical Data,* Macmillan Co., New York.

B. O. Peirce, *A Short Table of Integrals,* Ginn and Co., New York.

We shouldn't leave the subject without saying a word about numerical computation. So — for a word about numerical computation go to 401.

401

Sometimes it is not possible to find the integral of a function. However, it is still possible to find the area under a curve at least approximately. A simple approximate method is to plot the function on graph paper with a fine mesh and to count the number of squares in the area.

Another rough and ready method is to plot the curve on uniform heavy cardboard, cut out the desired area and weigh it. There are also devices called planimeters, which mechanically integrate any area whose boundary is traced by the planimeter.

It is always possible to evaluate a definite integral to any desired accuracy by numerical integration. To find the area under a curve, simply divide it into a convenient number of strips,

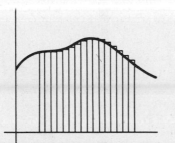

find the height of each, multiply by the width and add them together. The narrower the strips, the better will be the answer — but the more the work. However, the use of high speed computing machines makes this approach highly effective where in the past it was often not practical.

Go to frame 402.

402

Well, here you are at the very last frame. You should get some reward for all your effort — all we can do is promise that there is only one more *"Go To"* left in the book.

The next chapter is a review and lists in outline all the ideas presented in the book. Even though you may have already read part of that chapter you should now study it all. You may also find it is handy for future reference.

The Appendixes are crammed full of interesting tidbits: derivations of formulas, explanations of special topics, and the like.

In case you are a glutton for punishment or simply want a little more practice, there is a list of review problems, along with the answers, starting on page 275.

Now go to Chapter IV.

REVIEW

This chapter is a review and concise summary of what you have learned. Proofs and detailed explanations given in the preceding three chapters are not repeated here; instead, references are given to the appropriate frames. Unlike the rest of the book, this chapter has no questions so it can be read from beginning to end like an ordinary text, except that you may occasionally want to refer back to earlier discussions.

Review of Chapter I.   A FEW PRELIMINARIES

Section 1. FUNCTIONS (frames 1–13)

A set is a collection of objects—not necessarily material objects—described in such a way that we have no doubt as to whether a particular object does or does not belong to the set. A set may be described by listing its elements or by a rule.

If each element of a set A is associated with exactly one element of set B, then this association is called a *function* from A to B. The set A is called the domain of the function. (Some comments about an alternative definition of a function are given in Appendix B1, page 262.)

If a symbol, such as x, is used to represent any element of the set A (the domain of the function), it is called the *independent variable*. If the symbol y represents the element of the set B associated by the function with the element x, we call y the *dependent variable*.

One way to specify a function is to list in detail the association between all the corresponding elements of the two sets. Another way is to present a rule for finding the dependent variable in terms of the independent variable. Thus, for example, a function associating the independent variable t with the dependent variable S could be specified by

$$S = 2 t^2 + 6t.$$

Unless otherwise stated, we shall assume that the independent variable can take on the value of any real number for which the dependent variable is also a real number.

We usually represent a function by a letter such as f. If the independent variable is x, the dependent variable y associated by the function f is often written as $f(x)$, which is read "f of x." Other symbols can be used, such as $z = H(v)$.

Section 2. GRAPHS (frames 14–22)

A convenient way to represent a function is to plot a graph as described in frames 15–18. The mutually perpendicular *coordinate axes* intersect at the *origin*. The axis that runs horizontally is called the *horizontal axis* or *x-axis*. The axis that runs vertically is called the *vertical axis* or *y-axis*. The value of the x-coordinate of a point is called the *abscissa*, and the value of the y-coordinate is called the *ordinate*.

The *constant function* results from the association of a single fixed number, c, with *all* values of the independent variable, x. The absolute value function $|x|$ is defined by

$$|x| = x \quad \text{if } x \geq 0.$$

$$|x| = -x \quad \text{if } x < 0.$$

Section 3. LINEAR AND QUADRATIC FUNCTIONS
(frames 23–39)

An equation of the form $y = mx + b$ where m and b are constants is called *linear* because its graph is a straight line. The slope of a linear function is defined by

$$\text{slope} = \frac{y_2 - y_1}{x_2 - x_1} = \frac{y_1 - y_2}{x_1 - x_2}.$$

From the definition it is easy to see (frame 29) that the slope of the above linear equation is m.

An equation of the form $y = ax^2 + bx + c$ where a, b, and c are constants, is called a *quadratic equation*. Its graph is called a *parabola*. The values of x at $y = 0$ satisfy $ax^2 + bx + c = 0$ and are called the *roots* of the equation. Not all quadratic equations have real roots. The equation $ax^2 + bx + c = 0$ has two roots given by

$$x = \frac{-b \pm \sqrt{b^2 - 4ac}}{2a}$$

Section 4. TRIGONOMETRY (frames 40–73)

Angles are measured in either *degrees* or *radians*.

A circle is divided into 360 equal *degrees*. The number of *radians* in an angle is equal to the length of the subtending arc divided by the length of the radius (frame 42). The relation between degrees and radians is

$$1 \text{ rad} = \frac{360°}{2\pi}.$$

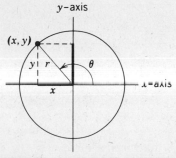

The trigonometric functions are defined in conjunction with the figure.

The definitions are

$$\sin \theta = \frac{y}{r} \qquad\qquad \cos \theta = \frac{x}{r}$$

$$\tan \theta = \frac{y}{x} \qquad\qquad \cot \theta = \frac{1}{\tan \theta} = \frac{x}{y}$$

$$\sec \theta = \frac{1}{\cos \theta} = \frac{r}{x} \qquad\qquad \csc \theta = \frac{1}{\sin \theta} = \frac{r}{y}.$$

Although $r = \sqrt{x^2 + y^2}$ is always positive, x and y can be either positive or negative and the above quantities may be positive or negative depending on the value of θ. From the Pythagorean Theorem it is easy to see (frame 56) that

$$\sin^2 \theta + \cos^2 \theta = 1.$$

The sines and cosines for the sum of two angles are given by:

$$\sin (\theta + \phi) = \sin \theta \cos \phi + \cos \theta \sin \phi$$

$$\cos (\theta + \phi) = \cos \theta \cos \phi - \sin \theta \sin \phi.$$

The inverse trigonometric function designates the angle for which the trigonometric function has the specified value. Thus the inverse trigonometric function to $y = \sin \theta$ is $\theta = \arcsin y$, which is read "arc sine of y" and stands for the angle whose sine is y. The arccos, arctan, etc., are similarly defined.

Section 4. EXPONENTIALS AND LOGARITHMS
(frames 74–96)

If a is multiplied by itself as $aaa...$ with m factors, the product is written as a^m. Furthermore, by definition $a^{-m} = 1/a^m$. From this it follows that

$$a^m a^n = a^{(m+n)}$$

$$a^m / a^n = a^{(m-n)}$$

$$a^0 = a^m / a^m = 1$$

$$(a^m)^n = a^{(mn)}$$

$$(ab)^m = a^m b^m.$$

If $b^n = a$, b is called the n'th root of a and is written as $b = a^{1/n}$. If m and n are integers

$$a^{m/n} = (a^{1/n})^m.$$

The meaning of exponents can be extended to irrational numbers (frame 84) and the above relations also apply with irrational exponents, so $(a^x)^b = a^{bx}$, etc.

The definition of log x (the logarithm of x to the base 10) is

$$x = 10^{\log x}$$

The following important relations can easily be seen to apply to logarithms (frame 91)

$$\log(ab) = \log(a) + \log(b)$$

$$\log(a/b) = \log(a) - \log(b)$$

$$\log(a^n) = n \log(a).$$

The logarithm of x to another base r is written as $\log_r x$ and is defined by

$$x = r^{\log_r x}.$$

The above three relations for logarithms of a and b are correct for logarithms to any base provided the same base is used for all the logarithms in each equation. Logarithms of x to two different bases, e and 10, can be related by

$$\log_e x = \frac{\log_{10} x}{\log_{10} e} = 2.303 \log_{10} x. \qquad \text{(frame 223)}$$

Review of Chapter II. DIFFERENTIAL CALCULUS

Section 1. LIMITS (frames 97–115)

Definition of a Limit: Let $f(x)$ be defined for all x in an interval about $x = a$, but not necessarily at $x = a$. If there is a number L such that to each positive number ϵ there corresponds a positive number δ such that

$$|f(x) - L| < \epsilon \qquad \text{provided} \qquad 0 < |x - a| < \delta$$

we say that L is the *limit* of $f(x)$ as x approaches a, and write

$$\lim_{x \to a} f(x) = L.$$

The ordinary algebraic manipulations can be performed with limits as shown in Appendix A2; thus

$$\lim_{x \to a} [F(x) + G(x)] = \lim_{x \to a} F(x) + \lim_{x \to a} G(x),$$

Two trigonometric limits are of particular interest (Appendix A3):

$$\lim_{\theta \to 0} \frac{\sin \theta}{\theta} = 1 \qquad \text{and} \qquad \lim_{\theta \to 0} \frac{1 - \cos \theta}{\theta} = 0.$$

The following limit is of such great interest in calculus that it is given the special name e, as discussed in frame 109 and Appendix A8:

$$e = \lim_{x \to 0} (1 + x)^{1/x} = 2.71828\ldots\ldots$$

Section 2. VELOCITY (frames 116–145)

If the function S represents the distance from a fixed location of a point moving at a varying speed along a straight line, the *average velocity*, $\overline{v}$, between times t_1 and t_2 is given by

$$\overline{v} = \frac{S_2 - S_1}{t_2 - t_1}$$

whereas the *instantaneous velocity, v,* (frame 133) at time t_1 is

$$v = \lim_{t_2 \to t_1} \frac{S_2 - S_1}{t_2 - t_1}.$$

This equals the slope at time t_1 of the curve of S plotted in terms of time (frame 131). It is often convenient to write $S_2 - S_1 = \Delta S$ and $t_2 - t_1 = \Delta t$, so

$$v = \lim_{\Delta t \to 0} \frac{\Delta S}{\Delta t}.$$

Section 3. DERIVATIVES (frames 146–159)

If $y = f(x)$, the rate of change of y with respect to x is

$\lim_{\Delta x \to 0} \frac{\Delta y}{\Delta x}$. The $\lim_{\Delta x \to 0} \frac{\Delta y}{\Delta x}$ is called the *derivative* of y with re-

spect to x and is written as $\frac{dy}{dx}$ (sometimes it is written y'). Thus

$$\frac{dy}{dx} = \lim_{\Delta x \to 0} \frac{\Delta y}{\Delta x} = \lim_{x_2 \to x_1} \frac{y_2 - y_1}{x_2 - x_1} = \lim_{x_2 \to x_1} \frac{f(x_2) - f(x_1)}{x_2 - x_1}$$

is the derivative of y with respect to x. The derivative $\frac{dy}{dx}$ is

equal to the slope of the curve of y plotted against x.

Section 4. GRAPHS OF FUNCTIONS AND THEIR DERIVATIVES (frames 160–169)

From a graph of a function we can obtain the slope of the curve at different points and by plotting a new curve of the slopes we can determine the general character and qualitative behavior of the derivative. See Section 4 for examples.

Sections 5–8. DIFFERENTIATION (frames 170–244)

From the definition of the derivative a number of formulae for differentiation can be derived. We will review just one example here; the method is typical. Let u and v be variables that depend on x.

$$\frac{d(uv)}{dx} = \lim_{\Delta x \to 0} \frac{\Delta(uv)}{\Delta x} = \lim_{\Delta x \to 0} \frac{(u + \Delta u)(v + \Delta v) - uv}{\Delta x}$$

$$\frac{d(uv)}{dx} = \lim_{\Delta x \to 0} \frac{uv + u\Delta v + v\Delta u + \Delta u \Delta v - uv}{\Delta x}$$

$$= u \lim_{\Delta x \to 0} \frac{\Delta v}{\Delta x} + v \lim_{\Delta x \to 0} \frac{\Delta u}{\Delta x} + \lim_{\Delta x \to 0} \frac{\Delta u \Delta v}{\Delta x}$$

$$= u \frac{dv}{dx} + v \frac{du}{dx} + 0.$$

The important relations which you should remember are listed here. There is a more complete list in Table 1, page 283. In the following expressions u and v are variables that depend on x, w depends on u, which in turn depends on x, and a and n are constants. All angles are measured in radians.

(frame)

$$\frac{da}{dx} = 0 \qquad\qquad 172$$

$$\frac{d}{dx}(ax) = a \qquad\qquad 174$$

$$\frac{dx^n}{dx} = nx^{n-1} \qquad\qquad 180$$

$$\frac{d}{dx}(u + v) = \frac{du}{dx} + \frac{dv}{dx} \qquad\qquad 186$$

$$\frac{d}{dx}(uv) = u\frac{dv}{dx} + v\frac{du}{dx} \qquad\qquad 189$$

$$\frac{d}{dx}\left(\frac{u}{v}\right) = \frac{1}{v^2}\left[v\frac{du}{dx} - u\frac{dv}{dx}\right] \qquad\qquad 202$$

$$\frac{dw}{dx} = \frac{dw}{du}\frac{du}{dx} \qquad\qquad 194$$

$$\frac{d \sin x}{dx} = \cos x \qquad\qquad 210$$

$$\frac{d \cos x}{dx} = -\sin x \qquad\qquad 211$$

$$\frac{d \ln x}{dx} = \frac{1}{x}$$

$$\frac{de^x}{dx} = e^x$$

In the above list $e = 2.71828 \ldots$. and $\ln x$ is the natural logarithm of x defined by $\ln x = \log_e x$.

More complicated functions can ordinarily be differentiated by applying several of the rules in Table 1 successively. Thus

$$\frac{d}{dx}(x^3 + 3x^2 \sin 2x) = \frac{dx^3}{dx} + 3\frac{dx^2}{dx}\sin 2x + 3x^2\frac{d\sin 2x}{dx}$$

$$= 3x^2 + 6x \sin 2x + 3x^2\frac{d\sin 2x}{d(2x)}\frac{d(2x)}{dx}$$

$$= 3x^2 + 6x \sin 2x + 6x^2 \cos 2x.$$

Section 9. HIGHER ORDER DERIVATIVES (frames 245–252)

If we differentiate $\frac{dy}{dx}$ with respect to x, the result is called the *second derivative* of y with respect to x and is written $\frac{d^2y}{dx^2}$. Likewise the n'th derivative of y with respect to x is the result of differentiating y n times successively with respect to x and is written $\frac{d^ny}{dx^n}$.

Section 10. MAXIMA AND MINIMA (frames 253–264)

If $f(x)$ has a maximum or minimum value for some value of x, then its derivative $\frac{df}{dx}$ is zero for that x. (Restrictions to this statement are given on p. 144). If in addition $\frac{d^2f}{dx^2} < 0$, $f(x)$ has maximum value. If on the other hand $\frac{d^2f}{dx^2} > 0$, $f(x)$, has a minimum value there.

Section 11. DIFFERENTIALS (frames 265–275)

If x is an independent variable and $y = f(x)$, the *differential* dx of x is defined as equal to any increment, $x_2 - x_1$, where x_1 is the point of interest. The differential dx can be positive or negative, large or small, as we please. Then dx, like x, is an independent variable.   The differential dy is then *defined* by the following rule

$$dy = (\frac{dy}{dx})\, dx.$$

Although the meaning of the derivative, $\frac{dy}{dx}$, is $\lim\limits_{\Delta x \to 0} \frac{\Delta y}{\Delta x}$, we see that it can now be interpreted as the ratio of the differentials dy and dx.   As discussed in frames 268 and 269, dy is not the same as Δy, though

$$\lim_{dx=\Delta x \to 0} \frac{dy}{\Delta y} = 1.$$

The differentiation formulae can easily be written in terms of differentials. Thus if $y = x^n$

$$dy = d(x^n) = \frac{d(x^n)}{dx}\, dx = nx^{n-1}\, dx.$$

A useful relation which is implied by the differential notation and discussed further in Appendix A9 is

$$\frac{dx}{dy} = 1/(\frac{dy}{dx}).$$

Review of Chapter III. INTEGRAL CALCULUS

Section 1. THE INDEFINITE INTEGRAL (frames 289–301)

Suppose $\frac{dF(x)}{dx} = f(x).$

Then $F(x)$ is called the *indefinite integral* of $f(x)$.   This statement is written symbolically in the form

$$F(x) = \int f(x)\, dx.$$

The equation is read "$F(x)$ equals the *indefinite integral* of $f(x)$." The function, $f(x)$, which is integrated is called the *integrand*.   Since the derivative of a constant is zero, any arbitrary constant c can be added to an indefinite integral and the sum will also be an indefinite integral of the same function $f(x)$. Furthermore, any two indefinite integrals of a given function can differ only by a constant (frame 296 and Appendix A12).

Section 2. INTEGRATION (frames 302–325)

Indefinite integrals are often found by hunting for an expression which, when differentiated, gives the integrand. Thus from the earlier result that

$$\frac{d \cos x}{dx} = -\sin x$$

we have that

$$\int \sin x \, dx = -\cos x + c.$$

By starting with known derivatives as in Table 1, a useful list of integrals can be found. Such a list is given in frame 307 and for convenience is repeated in Table 2 on page 285. You can reconstruct the most important of these formulas from the differentiation expressions in Table 1. More complicated integrals can often be found in large tables, such as those listed in the references on page 274.

Often it is necessary to use several different procedures to obtain an integral. As an example we use both the change of variable procedure associated with formula 19 of the table and formula 10 for integrating the sine in the following example (frame 313):

$$\int \sin 3 \, x \, dx = \frac{1}{3} \int \sin 3x \, d(3x) = -\frac{1}{3} \cos 3x + c.$$

Section 3. THE AREA UNDER A CURVE (frames 326–346)

Let $A(x)$ be the area between the curve $f(x)$ and the x-axis bounded by vertical lines crossing the x-axis at a and x. We can then see that

$$\frac{d \, A(x)}{dx} = f(x). \qquad \text{(frame 333)}$$

If $F(x)$ is an indefinite integral of $f(x)$ so

$$F(x) = \int f(x) \, dx,$$

then (frames 340 and 344)

$$A(x) = F(x) - F(a) = F(x) \Big|_a^x = \int f(x) \, dx \Big|_a^x$$

where by definition $F(x) \Big|_a^b = F(b) - F(a)$.

Section 4. DEFINITE INTEGRALS (frames 347–365)

An alternative expression for the area A under a curve $f(x)$ between $x = a$ and $x = b$ can be found by dividing the area into narrow strips parallel to the y-axis, each of area $f(x_i) \Delta x$, and summing the strips. In the limit as the width of each strip approaches zero the limit of the sum approaches the area under the curve. Thus (frame 352),

$$A = \lim_{\Delta x \to 0} \sum_{i=1}^{n} f(x_i) \, \Delta x.$$

Such a limit is so important that it is given a special name and symbol. It is called the *definite integral* and is written $\int_a^b f(x) \, dx$. Hence by definition

$$\int_a^b f(x) \, dx = \lim_{\Delta x \to 0} \sum f(x_i) \, \Delta x.$$

As a result of this discussion, we see that

$$A = \int_a^b f(x) \, dx.$$

However, we have seen that the area can also be evaluated in terms of the *indefinite integral*,

$$F(x) = \int f(x) \, dx,$$

by

$$A = F(b) - F(a) = F(x) \Big|_a^b = \int f(x) \, dx \Big|_a^b.$$

Therefore, by equating the two expressions for A we have the general evaluation of the *definite* integral in terms of the *indefinite* integral.

$$\int_a^b f(x) \, dx = F(x) \Big|_a^b = \int f(x) \, dx \Big|_a^b.$$

This result is often called the Fundamental Theorem of Integral Calculus.

Section 5. SOME APPLICATIONS OF INTEGRATION
            (frames 366–381)

If we know $v(t)$, the *velocity* of a particle as a function of t, we can obtain the *position* of the particle as a function of time by integration. We saw earlier

$$v = \frac{dS}{dt}$$

so

$$dS = v\,dt$$

and if we integrate both sides of the equation from the initial point ($t = t_0$, $S = 0$) to the final point (t, S), we have

$$S = \int_{t_0}^{t} v\,dt.$$

Section 6. MULTIPLE INTEGRALS (frames 382–399)

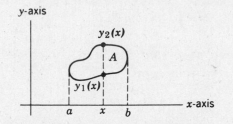

Consider the indicated area enclosed by the curve for which y depends on x as $y_2(x)$ at the top of the figure as $y_1(x)$ at the bottom. Then the enclosed area A is given by

$$A = \lim_{\Delta x \to 0} \Sigma_x \left\{ \lim_{\Delta y \to 0} \Sigma_y \, \Delta y \right\} \Delta x$$

$$= \int_a^b \left[\int_{y_1}^{y_2} dy \right] dx.$$

An integral of the above form is called a *double integral*, which is a particular example of a *multiple integral*. In the evaluation of multiple integrals special care must be taken to use the correct expression for the limits. Thus y_1 and y_2, the limits for the y integration, are the maximum and minimum values of y

for a particular x.   As a result y_1 and y_2 in general depend on x and consequently they contribute to the integrand of the next integration over x.

In a similar fashion (frame 390) if $g(x,y)$ is a variable that depends on both x and y we can evaluate a multiple integral in the form

$$G = \int_a^b \left[\int_{y_1}^{y_2} g(x,\,y)\;dy \right] dx.$$

The procedure is straightforward: the limits on each of the integrals are found, and the y integration is performed with x treated like a constant in $g(x,y)$.   Next, the x integration is done. The procedure is easily extended to integrals involving even more variables.

Section 7. CONCLUSION (frames 400–402)

You are now finished.   Congratulations!   You don't need to do any more work to complete this book.   However, if you skipped some of the proofs in Appendix A we suggest you read them now. You may also want to study some of the additional topics that are described in Appendix B.   Finally, if you would like to have some more practice, you should try some of the review problems starting on p 275.

Good luck!

APPENDIX A

DERIVATIONS

In this appendix derivations are given of certain of the formulas and theorems not derived earlier.

Appendix A1

TRIGONOMETRIC FUNCTIONS OF SUMS OF ANGLES

A formula can easily be derived for the sine of the sum of two angles, θ and ϕ, with the aid of the drawing in which the radius of the circle is unity.

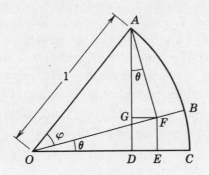

$$\sin (\theta + \phi) = AD = FE + AG$$

$$= OF \sin \theta + AF \cos \theta$$

$$= \sin \theta \cos \phi + \cos \theta \sin \phi$$

In a similar fashion with the same figure,

$$\cos (\theta + \phi) = OD = OE - DE$$

$$= OF \cos \theta - AF \sin \theta$$

$$= \cos \theta \cos \phi - \sin \theta \sin \phi .$$

Appendix A2

SOME THEOREMS ON LIMITS

In this appendix we shall prove several useful theorems on limits. These theorems will show that the usual algebraic manipulations can be done with expressions involving limits. We shall show for example that

$$\lim_{x \to a} [F(x) + G(x)] = \lim_{x \to a} F(x) + \lim_{x \to a} G(x).$$

Although such results are intuitively reasonable they require a formal proof.

Before deriving these theorems, we need to note some general properties of the absolute value function introduced in frame 20, page 13. These properties are

$$|a + b| \le |a| + |b| \tag{1}$$

$$|ab| = |a| \times |b|. \tag{2}$$

It is easy to see that these relations are true by considering in turn each of all the possible cases: a and b both negative, both positive, of opposite sign, and one or both equal to zero.

We are now ready to discuss theorems on limits which apply to any two functions F and G such that

$$\lim_{x \to a} F(x) = L \quad \text{and} \quad \lim_{x \to a} G(x) = M.$$

Theorem 1

$$\lim_{x \to a} [F(x) + G(x)] = \lim_{x \to a} F(x) + \lim_{x \to a} G(x).$$

Proof: By Equation (1)

$$|F(x) + G(x) - (L + M)| = |[F(x) - L] + [G(x) - M]|$$
$$\leq |F(x) - L| + |G(x) - M|.$$

Using the definition of the limit (frame 105, p. 60) we see that for any positive number ϵ we can find a positive number δ such that

$$|F(x) - L| < \frac{\epsilon}{2} \quad \text{and} \quad |G(x) - M| < \frac{\epsilon}{2}$$

provided $0 < |x - a| < \delta$. (At first sight this may appear to differ from the definition of the limit since the symbol ϵ instead of $\epsilon/2$ appeared there.   However, the statements apply for any positive number and $\epsilon/2$ is also a positive number.)

The above equations may be combined to give

$$|F(x) + G(x) - (L + M)| < \frac{\epsilon}{2} + \frac{\epsilon}{2} = \epsilon.$$

Therefore, by the definition of the limit in frame 105, page 60

$$\lim_{x \to a} [F(x) + G(x)] = L + M = \lim_{x \to a} F(x) + \lim_{x \to a} G(x).$$

Theorem 2

$$\lim_{x \to a} [F(x) G(x)] = [\lim_{x \to a} F(x)] [\lim_{x \to a} G(x)].$$

Proof: The proof is somewhat similar to the preceding. By writing out all the terms we can see that the following is true identically:

$$F(x) G(x) - LM = [F(x) - L] [G(x) - M] + L [G(x) - M]$$
$$+ M [F(x) - L].$$

Therefore, by Equation (1)

$$| F(x) G(x) - LM | \leq |[F(x) - L] [G(x) - M]| + | L [G(x) - M]|$$
$$+ | M [F(x) - L]|.$$

Let ϵ be any positive number less than 1. Then by the meaning of limits we can find a positive number δ such that if $0 < | x - a | < \delta$

$$| F(x) - L | < \frac{\epsilon}{2}, \quad | L [G(x) - M]| < \frac{\epsilon}{4}, \quad | M [F(x) - L]| < \frac{\epsilon}{4},$$

$$\text{and } | G(x) - M | < \frac{\epsilon}{2}.$$

Then

$$| F(x) G(x) - LM | < \frac{\epsilon^2}{4} + \frac{\epsilon}{4} + \frac{\epsilon}{4} = \frac{\epsilon^2}{4} + \frac{\epsilon}{2} \leq \frac{\epsilon}{4} + \frac{\epsilon}{2} = \frac{3}{4} \epsilon$$

where the next to the last step arises as a result of our earlier restriction to $\epsilon < 1$.

Consequently

$$| F(x) G(x) - LM | < \epsilon$$

so by the definition of the limit

$$\lim_{x \to a} [F(x) G(x)] = LM = [\lim_{x \to a} F(x)] [\lim_{x \to a} G(x)].$$

Theorem 3

$$\lim_{x \to a} \frac{F(x)}{G(x)} = \frac{\lim\limits_{x \to a} F(x)}{\lim\limits_{x \to a} G(x)} \qquad \text{provided } \lim_{x \to a} G(x) \neq 0.$$

Proof: Since $\lim\limits_{x \to a} G(x) \neq 0$, we can select a value of δ suffi-
ciently small that $G(x) \neq 0$ for $0 < |x - a| < \delta$. Then we can write

$$\lim_{x \to a} F(x) = \lim_{x \to a} \left[G(x) \frac{F(x)}{G(x)} \right] = \lim_{x \to a} G(x) \lim_{x \to a} \frac{F(x)}{G(x)}$$

$$= M \lim_{x \to a} \frac{F(x)}{G(x)}$$

where $M = \lim\limits_{x \to a} G(x)$.

Therefore, since $M \neq 0$, we have

$$\lim_{x \to a} \frac{F(x)}{G(x)} = \frac{\lim\limits_{x \to a} F(x)}{M} = \frac{\lim\limits_{x \to a} F(x)}{\lim\limits_{x \to a} G(x)}.$$

Appendix A3

Limits Involving Trigonometric Functions

1. Proof that

$$\lim_{\theta \to 0} \frac{\sin \theta}{\theta} = 1.$$

To prove this draw an arc of a unit circle as shown, such that $AB = AE = 1$ and $\theta = \angle EAB$. Geometrically it is apparent that

area $ADE \geq$ area $ABE \geq ABC$

Therefore $\frac{1}{2}(\overline{AE})(\overline{DE}) \geq$ area $ABE \geq \frac{1}{2}(\overline{AC})(\overline{BC})$. (The symbol $\overline{AE}$ represents the length of the straight line segment between A and E.)

Since the area of the circle is π, we have

$$\text{area } ABE = \pi \frac{\theta}{2\pi} = \frac{1}{2}\theta.$$

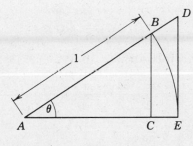

Using the fact that $\overline{DE} = \tan \theta$ we obtain

$$\frac{1}{2}\tan \theta \geq \frac{1}{2}\theta \geq \frac{1}{2}\cos \theta \sin \theta.$$

Dividing through by $\frac{1}{2}\sin \theta$ yields

$$\frac{1}{\cos \theta} \geq \frac{\theta}{\sin \theta} \geq \cos \theta.$$

Take the reciprocals of this expression. Since the reciprocal of a large number is smaller than the reciprocal of a small number, (providing both numbers are positive), this operation reverses the order of the inequality:

$$\cos \theta \leq \frac{\sin \theta}{\theta} \leq \frac{1}{\cos \theta}$$

So

$$\lim_{\theta \to 0} \cos \theta \leq \lim_{\theta \to 0} \frac{\sin \theta}{\theta} \leq \lim_{\theta \to 0} \frac{1}{\cos \theta}$$

and

$$1 \le \lim_{\theta \to 0} \frac{\sin \theta}{\theta} \le 1.$$

Therefore

$$\lim_{\theta \to 0} \frac{\sin \theta}{\theta} = 1.$$

2. Proof that

$$\lim_{\theta \to 0} \frac{1 - \cos \theta}{\theta} = 0.$$

This can be proved as follows:

$$1 - \cos \theta = \frac{(1 - \cos \theta)(1 + \cos \theta)}{1 + \cos \theta} = \frac{1 - \cos^2 \theta}{1 + \cos \theta}$$

$$= \frac{\sin^2 \theta}{1 + \cos \theta} \le \sin^2 \theta \text{ for } 0 \le \theta < \frac{\pi}{2}.$$

Therefore in this limit

$$\frac{1 - \cos \theta}{\theta} \le \frac{\sin^2 \theta}{\theta}.$$

We then have

$$\lim_{\theta \to 0} \frac{1 - \cos \theta}{\theta} \le \left[\lim_{\theta \to 0} \frac{\sin \theta}{\theta} \right] \left[\lim_{\theta \to 0} \sin \theta \right] = 1 \times 0 = 0.$$

But for all positive θ, $0 \le \dfrac{1 - \cos \theta}{\theta}$.   Hence

$$0 \le \lim_{\theta \to 0} \frac{1 - \cos \theta}{\theta} \le 0.$$

The only way to satisfy both of these conditions is for

$$\lim_{\theta \to 0} \frac{1 - \cos \theta}{\theta} = 0.$$

Appendix A4

DIFFERENTIATION OF x^n

Consider first the case of n a positive integer.

$$y = x^n \tag{1}$$

$$y + \Delta y = (x + \Delta x)^n.$$

The right side can be expanded by the binomial theorem (if you are not familiar with this, look it up in any good algebra text) to give

$$y + \Delta y = (x + \Delta x)^n = x^n + nx^{n-1} \, \Delta x + \frac{n(n-1)}{1 \cdot 2} \, x^{n-2} \, \Delta x^2$$

$$+ \ldots\ldots + \Delta x^n. \tag{2}$$

If we subtract Equation (1) from Equation (2) and divide by Δx we have

$$\frac{\Delta y}{\Delta x} = nx^{n-1} + \frac{n(n-1)}{1 \cdot 2} \, x^{n-2} \, \Delta x + \ldots\ldots + \Delta x^{n-1}.$$

Therefore

$$\frac{dy}{dx} = \lim_{\Delta x \to 0} \frac{\Delta y}{\Delta x} = nx^{n-1}.$$

Although the above theorem has been proved only for n being a positive integer, we can also show it is true for $n = 1/q$ where q is a positive integer. Let

$$y = x^{1/q}$$

so

$$x = y^q.$$

By the preceding theorem, then

$$\frac{dx}{dy} = q \, y^{q-1}.$$

But by Appendix A11

$$\frac{dy}{dx} = 1/(\frac{dx}{dy}) = 1/[q \, y^{q-1}] = \frac{1}{q} y^{1-q} = \frac{1}{q} [x^{1/q}]^{(1-q)}$$

$$\frac{dy}{dx} = \frac{1}{q} x^{(1/q)-1} = nx^{n-1}.$$

We can further see that this theorem holds for $n = \frac{p}{q}$ where p and q are both positive integers.

$$y = x^n = x^{p/q}.$$

Let

$$w = x^{1/q}$$

so

$$y = w^p.$$

Then

$$\frac{dy}{dx} = \frac{dy}{dw}\frac{dw}{dx} = pw^{p-1} (\frac{1}{q}) x^{(1/q)-1} = px^{(p/q)-(1/q)} (\frac{1}{q}) x^{(1/q)-1}$$

$$= (\frac{p}{q}) x^{(p/q)-1} = nx^{n-1}.$$

So far we have seen that the rule for differentiating x^n applies if n is any positive fraction. We will now see that it applies for negative fractions as well. Let $n = -m$ where m is a positive fraction. Then

$$\frac{d(x^n)}{dx} = \frac{d(x^{-m})}{dx} = \frac{d}{dx} (\frac{1}{x^m}) = -\frac{dx^m/dx}{(x^m)^2}$$

$$= -\frac{mx^{m-1}}{x^{2m}} = (-m) x^{-m-1} = nx^{n-1}.$$

Up to now our discussion applies if n is any rational number. However, the result may be extended to any irrational real number by the method used in Frame 84, since an irrational number can be approximated as closely as desired by a fraction. Therefore, for any real number n, whether rational or irrational, and regardless of sign,

$$\frac{dx^n}{dx} = nx^{n-1}.$$

Appendix A5

DIFFERENTIATION OF TRIGONOMETRIC FUNCTIONS

From Appendix A1

$$\frac{d(\sin \theta)}{d\theta} = \lim_{\Delta\theta \to 0} \frac{\sin(\theta + \Delta\theta) - \sin \theta}{\Delta\theta}$$

$$= \lim_{\Delta\theta \to 0} \frac{\sin \theta \cos \Delta\theta + \cos \theta \sin \Delta\theta - \sin \theta}{\Delta\theta}$$

$$= \sin \theta \lim_{\Delta\theta \to 0} \frac{\cos \Delta\theta - 1}{\Delta\theta} + \cos \theta \lim_{\Delta\theta \to 0} \frac{\sin \Delta\theta}{\Delta\theta}.$$

The two limits were evaluated in Appendix A3 as 0 and 1 respectively, so

$$\frac{d(\sin \theta)}{d\theta} = \cos \theta.$$

Likewise,

$$\frac{d(\cos \theta)}{d\theta} = \lim_{\Delta\theta \to 0} \frac{\cos(\theta + \Delta\theta) - \cos \theta}{\Delta\theta}$$

$$= \lim_{\Delta\theta \to 0} \frac{\cos \theta \cos \Delta\theta - \sin \theta \sin \Delta\theta - \cos \theta}{\Delta\theta}$$

$$= \cos \theta \lim_{\Delta\theta \to 0} \frac{\cos \Delta\theta - 1}{\Delta\theta} - \sin \theta \lim_{\Delta\theta \to 0} \frac{\sin \Delta\theta}{\Delta\theta}$$

$$= -\sin \theta.$$

Derivatives of other trigonometric functions can be found by expressing them in terms of sines and cosines, as in Chapter II.

Appendix A6

DIFFERENTIATION OF THE PRODUCT OF TWO FUNCTIONS

Let $y = uv$ where u and v are variables which depend on x. Then

$$y + \Delta y = (u + \Delta u)(v + \Delta v) = uv + u\Delta v + v\Delta u + \Delta u \Delta v.$$

Then

$$\frac{dy}{dx} = \lim_{\Delta x \to 0} \frac{(y + \Delta y) - y}{\Delta x} = \lim_{\Delta x \to 0} \frac{(uv + u\Delta v + v\Delta u + \Delta u \Delta v) - uv}{\Delta x}$$

$$= \lim_{\Delta x \to 0} [u \frac{\Delta v}{\Delta x} + v \frac{\Delta u}{\Delta x} + \Delta u \frac{\Delta v}{\Delta x}].$$

But

$$\lim_{\Delta x \to 0} \Delta u \frac{\Delta v}{\Delta x} = [\lim_{\Delta x \to 0} \Delta u] \times [\lim_{\Delta x \to 0} \frac{\Delta v}{\Delta x}] = 0 \times \frac{dv}{dx} = 0,$$

where we have used theorem 2 of Appendix A2. Thus

$$\frac{dy}{dx} = u \lim_{\Delta x \to 0} \frac{\Delta v}{\Delta x} + v \lim_{\Delta x \to 0} \frac{\Delta u}{\Delta x} = u \frac{dv}{dx} + v \frac{du}{dx}.$$

Appendix A7

CHAIN RULE FOR DIFFERENTIATION

Let $w(u)$ depend on u, which in turn depends on x.   Then

$$\Delta w = w(u + \Delta u) - w(u)$$

so

$$\frac{\Delta w}{\Delta x} = \frac{\Delta w}{\Delta u} \frac{\Delta u}{\Delta x} = \frac{w(u + \Delta u) - w(u)}{\Delta u} \frac{\Delta u}{\Delta x}.$$

Therefore, using Theorem 2 of Appendix A1, we have

$$\frac{dw}{dx} = \lim_{\Delta x \to 0} \frac{\Delta w}{\Delta x} = \lim_{\Delta x \to 0} \frac{\Delta w}{\Delta u} \lim_{\Delta x \to 0} \frac{\Delta u}{\Delta x} = (\frac{dw}{du})(\frac{du}{dx}).$$

Appendix A8

THE QUANTITY e

In frame 109 we introduced the quantity $e = 2.71828 \ldots \ldots$.
Here we will consider the exact definition of e by which the
above number can be extended to as many figures as desired. The
exact definition of e is in terms of a limit and is

$$e = \lim_{l \to 0} (1 + l)^{1/l}.$$

The convergent character of the above expression can be seen
by considering the following table of values of $(1 + l)^{1/l}$ for
successively smaller values of l.   As an exercise you should
check the validity of the calculation for the first three values
of l in the table.

l	$(1 + l)^{1/l}$
1	2
1/2	2.25
1/3	2.37
1/10	2.59
1/100	2.70
1/1000	2.72
1/10,000	2.72

With digital computing machines the value of e has been calculated to thousands of decimal places. The value 2.71828 is sufficiently accurate for most normal purposes. In Appendix A9 you
will see why e defined in the above way is so important.

Appendix A9

DIFFERENTIATION OF ln x

Let

$$y = \ln x$$

$$y + \Delta y = \ln (x + \Delta x).$$

Then

$$\frac{\Delta y}{\Delta x} = \frac{y + \Delta y - y}{\Delta x} = \frac{\ln (x + \Delta x) - \ln x}{\Delta x}.$$

From frame 91,

$$\frac{\Delta y}{\Delta x} = \frac{1}{\Delta x} \ln \left(\frac{x + \Delta x}{x}\right) = \frac{1}{x} \frac{x}{\Delta x} \ln \left(1 + \frac{\Delta x}{x}\right)$$

$$= \frac{1}{x} \ln \left(1 + \frac{\Delta x}{x}\right)^{x/\Delta x} = \frac{1}{x} \ln (1 + l)^{1/l}$$

where we have written l for $\dfrac{\Delta x}{x}$. Note that as $\Delta x \to 0$, $l \to 0$.

Therefore,

$$\frac{dy}{dx} = \lim_{\Delta x \to 0} \frac{\Delta y}{\Delta x} = \lim_{\Delta x \to 0} \left[\frac{1}{x} \ln (1 + l)^{1/l}\right]$$

$$= \frac{1}{x} \ln \left[\lim_{l \to 0} (1 + l)^{1/l}\right]$$

$$= \frac{1}{x} \ln e = \frac{1}{x}$$

since $\ln e = \log_e e = 1$.

Appendix A10

DIFFERENTIALS WHEN BOTH VARIABLES DEPEND ON A THIRD VARIABLE

The relation $dw = (\dfrac{dw}{du})\, du$

is true even when both w and u depend on a third variable. To prove this let both u and w depend on x. Then

$$dw = (\frac{dw}{dx})\, dx \text{ and } du = (\frac{du}{dx})\, dx. \tag{1}$$

By the chain rule for differentiating,

$$(\frac{dw}{dx}) = (\frac{dw}{du})\,(\frac{du}{dx})$$

and multiplying through by dx we have

$$(\frac{dw}{dx})\, dx = (\frac{dw}{du})\,(\frac{du}{dx})\, dx,$$

so by Equation (1)

$$dw = \frac{dw}{du}\, du.$$

This theorem justifies the use of the differential notation since it shows that with the differential notation the *chain rule* takes the form of an algebraic identity

$$\frac{dw}{dx} = \frac{dw}{du}\frac{du}{dx}.$$

Appendix A11

PROOF THAT $\dfrac{dy}{dx} = 1 / \dfrac{dx}{dy}$

If a function is specified by an equation $y = f(x)$ it is ordinarily possible, for at least limited intervals of x, to reverse the roles of the dependent and independent variables and to allow the equation to determine the value of x for a given value of y. (This can not always be done as in the case of the equation $y = a$, where a is a constant.) When such an inversion is possible the two derivatives are related by

$$\frac{dy}{dx} = 1 / \frac{dx}{dy}.$$

The relation can be seen as follows:

$$\frac{dy}{dx} = \lim_{\Delta x \to 0} \frac{\Delta y}{\Delta x} = \lim_{\Delta x \to 0} \frac{1}{(\Delta x / \Delta y)} = \frac{1}{\displaystyle\lim_{\Delta x \to 0} (\Delta x / \Delta y)}$$

by the limit theorems of Appendix 5. Furthermore, if $\displaystyle\lim_{\Delta x \to 0} \frac{\Delta y}{\Delta x} \neq 0$, then $\Delta y \to 0$ as $\Delta x \to 0$, so

$$\frac{dy}{dx} = \frac{1}{\displaystyle\lim_{\Delta y \to 0} (\Delta x / \Delta y)} = \frac{1}{(dx/dy)}.$$

This result is a further justification of the use of differential notation since normal arithmetic manipulation with differential notation immediately gives

$$\frac{dy}{dx} = 1 / \frac{dx}{dy}.$$

Appendix A12

PROOF THAT IF TWO FUNCTIONS HAVE THE SAME DERIVATIVE THEY DIFFER ONLY BY A CONSTANT

Let the functions be f and g.

Then

$$\frac{d f(x)}{dx} = \frac{d g(x)}{dx}$$

so

$$\frac{d}{dx}[f(x) - g(x)] = 0$$

Therefore

$$f(x) - g(x) = C$$

where C is a constant.

This proof depends on the assumption that if $\frac{d h(x)}{dx} = 0$ then $h(x)$ is a constant. This is indeed very plausible since the graph of the function $h(x)$ must always have zero slope and hence it should be a straight line parallel to the origin, i.e., $h(x) = C$. A more complicated analytic proof of this theorem is given in advanced books on calculus.

APPENDIX B

ADDITIONAL TOPICS

This appendix gives brief discussions of some additional topics in calculus.

Appendix B1

ALTERNATIVE DEFINITION OF FUNCTION

The definition of the word *function* we gave in frame 6 is that most frequently used in modern mathematics. However, some mathematicians and many other scientists continue to follow an earlier usage according to which the phrase *function of x* is defined instead of *function*. Here is one form of the older definition: If to every value of x in a certain set there corresponds a value of y in another set, then y is said to be a *function of x*. In other words, the dependent variable is a *function* of the independent variable whereas in frame 6 the function is the association from the independent variable to the dependent variable. If the temperature is known at different times, one would say with the older definition that the temperature is a function of the time. Likewise $y = f(x)$ in addition to being read as "y equals f of x" (as in frame 12) could be read with the older definition as "y *is* a function of x". Although the two statements sound almost identical, the difference reflects different points of view. For instance, if $f(x) = x^2$, with the older definition one might say that the function of x varies between 0 and 4 as x varies between 0 and 2. However, with the definition we have used the function is an association, and this does not change. What does change is the value of the dependent variable as the independent variable changes. That is, $f(x)$ varies, but the function does not.

In your future use of calculus you may have occasion to read books which follow either the earlier or the more recent meaning of the word "function". This should cause you no difficulty as long as you know of the existence of the two definitions, since the correct choice of meaning is usually quite clear from the context.

Appendix B2

PARTIAL DERIVATIVES

In this book we have almost exclusively considered functions defined for a single independent variable. Often, however, two or more independent variables are required to define the function; in this case we have to modify the idea of a derivative. As a simple example, suppose we consider the area of a rectangle, A, which is the product of its width, w, and length, l. Thus, $A = f(l,\ w)$ (read "f of l and w"), where $f(l,\ w)$ is here $l \times w$. In this discussion we will let l and w vary independently, so they both can be treated as independent variables.

If one of the variables, say w, is temporarily kept constant, then A depends on a single variable, and the rate of change of A with respect to l is simply $\dfrac{dA}{dl}$. However, because A really depends on two variables, we must modify the definition of the derivative.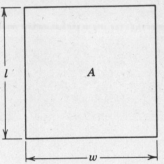

$$\begin{array}{cc} \text{The rate of change of } A \\ \text{with respect to } l \end{array} = \lim_{\Delta l \to 0} \frac{f(l + \Delta l,\ w) - f(l,\ w)}{\Delta l}$$

where it is understood that w is held constant as the limit is taken. The above quantity is called the *partial derivative* of A with respect to l and is written $\dfrac{\partial A}{\partial l}$. In other words the partial derivative is defined by

$$\frac{\partial A}{\partial l} = \frac{\partial f(l,\ w)}{\partial l} = \lim_{\Delta l \to 0} \frac{f(l + \Delta l,\ w) - f(l,\ w)}{\Delta l}.$$

In our example, $\dfrac{\partial A}{\partial l} = \lim_{\Delta l \to 0} \dfrac{(l + \Delta l) \times w - l \times w}{\Delta l} = w.$

Similarly, $\qquad \dfrac{\partial A}{\partial w} = \lim_{\Delta w \to 0} \dfrac{f(l,\ w + \Delta w) - f(l,\ w)}{\Delta w}$

$$= \lim_{\Delta w \to 0} \frac{l \times (w + \Delta w) - l \times w}{\Delta w} = l.$$

The differential of A due to changes in l and w of dl and dw, respectively, is by definition

$$dA = \frac{\partial A}{\partial l} \, dl + \frac{\partial A}{\partial w} \, dw.$$

By analogy with the argument in 269, it should be plausible that as $dl \to 0$ and $dw \to 0$, the increment in A, $\Delta A = f(l + \Delta l, \, w + \Delta w) - f(l, \, w)$, approaches dA.

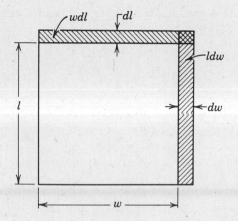

This result is shown by the figure. ΔA is the total increase in area due to dl and dw, and comprises all the shaded areas.

$$dA = \frac{\partial A}{\partial l} \, dl + \frac{\partial A}{\partial w} \, dw = w\,dl + l\,dw.$$

ΔA and dA differ by the area of the small rectangle in the upper right-hand corner. As $dl \to 0$, $dw \to 0$, the difference becomes negligible compared to the area of each strip.

The above discussion can be generalized to functions depending on any number of variables. For instance, let p depend on $q, r, s, \ldots$.

$$dp = \frac{\partial p}{\partial q} \, dq + \frac{\partial p}{\partial r} \, dr + \frac{\partial p}{\partial s} \, ds + \ldots.$$

Here is an example:

$$p = q^2 r \sin z$$

$$\frac{\partial p}{\partial q} = 2qr \sin z$$

$$\frac{\partial p}{\partial r} = q^2 \sin z$$

$$\frac{\partial p}{\partial z} = q^2 r \cos z$$

$$dp = 2qr \sin z \, dq + q^2 \sin z \, dr + q^2 r \cos z \, dz.$$

Here is another example:

The volume of a pyramid with height h and a rectangular base with dimensions l and w is

$$V = \frac{1}{3} lwh.$$

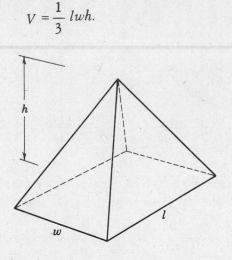

Thus

$$dV = \frac{1}{3} whdl + \frac{1}{3} lhdw + \frac{1}{3} lwdh.$$

If the dimensions are changed by small amounts dl, dw and dh, the volume changes by an amount $\Delta V \approx dV$, where dV is given by the expression above.

Appendix B3

IMPLICIT DIFFERENTIATION

Although most of the functions we use in this book can be written in the simple form $y = f(x)$, this is not always the case. Sometimes we have two variables related by an equation of the form $f(x, y) = 0$. [$f(x, y)$ means that the value of f depends on both x and y.] Here is an example: $x^2y + (y + x)^3 = 0$. We cannot easily solve this equation to yield a result of the form $y = g(x)$, or even $x = h(y)$. However, we can find $\dfrac{dy}{dx}$ by using the following procedure:

Let us differentiate both sides of the equation with respect to x remembering that y depends on x.

$$\frac{d}{dx}(x^2y) + \frac{d}{dx}(y + x)^3 = \frac{d}{dx}0 = 0$$

$$x^2 \frac{dy}{dx} + 2\,xy + 3\,(y + x)^2 \left[\frac{dy}{dx} + 1\right] = 0$$

$$\frac{dy}{dx}\left[(x^2 + 3\,(y + x)^2]\right] = -\,2xy - 3\,(y + x)^2$$

$$\frac{dy}{dx} = -\,\frac{2\,xy + 3\,(y + x)^2}{x^2 + 3\,(y + x)^2}.$$

A function defined by $f(x, y) = 0$ is called an *implicit* function since it implicitly determines the dependence of y on x (or, for that matter, the dependence of x on y in case we have to regard y as the independent variable). The process we have just used, differentiating each term of the equation $f(x, y) = 0$ with respect to the variable of interest, is called *implicit differentiation.*

Here is another example of implicit differentiation. Let $x^2 + y^2 = 1$. The problem is to find $\dfrac{dy}{dx}$. We will do this first by implicit differentiation, and then by solving the equation for y and using the normal procedure.

By differentiating both sides of the equation with respect to x, we obtain

$$2x + 2y\frac{dy}{dx} = 0.$$

Hence, $\dfrac{dy}{dx} = -\dfrac{2x}{2y} = -\dfrac{x}{y}$.

Alternatively, we can solve for y.

$$y^2 = 1 - x^2, \qquad y = \pm\sqrt{1 - x^2}$$

$$\frac{dy}{dx} = \pm\left(\frac{-2x}{\sqrt{1 - x^2}}\times\frac{1}{2}\right) = \mp\frac{x}{\sqrt{1 - x^2}} = -\frac{x}{y}.$$

We did not need to use implicit differentiation here since we could write the function in the form $y = f(x)$. Often, however, this cannot be done, as in the first example, and implicit differentiation is then necessary.

Appendix B4

DIFFERENTIATION OF THE INVERSE TRIGONOMETRIC FUNCTIONS

(1) Evaluation of $\dfrac{d}{dx}$ arcsin x.

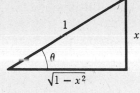

The angle θ is shown inscribed in a right triangle having unit hypotenuse, and an opposite side of length x. Therefore, $\sin \theta = x/1 = x$ and $\theta = \arcsin x$. Differentiating the first expression with respect to x yields

$$\frac{d \sin \theta}{dx} = 1.$$

Using the chain rule, we have

$$\frac{d}{dx} \sin \theta = \frac{d}{d\theta} \sin \theta \frac{d\theta}{dx} = \cos \theta \frac{d\theta}{dx} = 1.$$

Then,

$$\frac{d\theta}{dx} = \frac{d}{dx} \arcsin x = \frac{1}{\cos \theta}.$$

We can substitute the value $\cos \theta = \sqrt{1 - x^2}$ to obtain our final result:

$$\frac{d}{dx} \arcsin x = \frac{1}{\sqrt{1 - x^2}}.$$

Note that we must take the sign of $\sqrt{1 - x^2}$ to agree with that of $\cos \theta$.

(2) Evaluation of $\dfrac{d}{dx}$ arccos x.

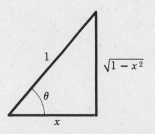

Using the triangle shown and above procedure we have

$$x = \cos \theta$$

$$\theta = \arccos x$$

$$\frac{d}{dx} \cos \theta = 1, \quad \frac{d}{d\theta} \cos \theta \frac{d\theta}{dx} = 1$$

$$\frac{d}{dx} \arccos x = \frac{-1}{\sin \theta} = \frac{-1}{\sqrt{1-x^2}} \;.$$

(3) Evaluation of $\dfrac{d}{dx}$ arctan x.

In the triangle shown, $\tan \theta = x$, so that $\theta = \arctan x$.

$$\frac{d}{dx} \tan \theta = \frac{d}{d\theta} \tan \theta \frac{d\theta}{dx} = 1. \quad \text{But } \frac{d}{d\theta} \tan \theta = \sec^2 \theta.$$

So $\dfrac{d\theta}{dx} = \dfrac{1}{\sec^2 \theta} = \cos^2 \theta = \dfrac{1}{1 + x^2}.$

$$\frac{d}{dx} \arctan x = \frac{1}{1 + x^2}.$$

(4) Evaluation of $\dfrac{d}{dx}$ arccot x.

Here $\cot \theta = x$, so that $\theta = $ arccot x.

$$\frac{d}{dx} \cot \theta = \frac{d}{d\theta} \cot \theta \frac{d\theta}{dx} = 1.$$

But

$$\frac{d}{d\theta} \cot \theta = - \csc^2 \theta$$

so

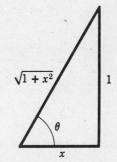

$$\frac{d\theta}{dx} = - \frac{1}{\csc^2 \theta} = - \sin^2 \theta = \frac{-1}{1 + x^2}.$$

$$\frac{d}{dx} \text{arccot } x = \frac{-1}{1 + x^2}.$$

Appendix B5

DIFFERENTIAL EQUATIONS

Any equation which involves a derivative of a function is called a *differential equation*. Such equations occur in many different applications of calculus, and their solution is the subject of a lively branch of mathematics. Here are two examples to show how a differential equation can occur.

1. THE GROWTH OF POPULATION.

Suppose we let n represent the number of people in a particular country. We assume that n is such a large number that we can neglect the fact that it must be an integer, and treat it as a continuous positive number. (In any application we would eventually round off n to the nearest integer.) The problem is this: Assume the birth rate is proportional to the population so nA children are born every year for every n people. A is the constant of proportionality. If the initial population of the country is n_0 people, how many people are there at some later time, T? (In this simple problem we will neglect deaths.)

If there are n people the total number of children born per year is nA. This is the *rate* of increase of population. That is,

$$\frac{dn}{dt} = nA.$$

The above differential equation is a particularly simple one. We can solve it by integration in the following manner:

$$\frac{dn}{n} = A \, dt.$$

Let us take the definite integral of both sides of the above equation. Initially we have $t = 0$ and $n = n_0$, and finally $t = T$, and $n = n(T)$. Thus

$$\int_{n_0}^{n(T)} \frac{dn}{n} = \int_0^T A \, dt.$$

The integral on the left should be familiar [if not, see Table 2, (5)]. Evaluating both integrals, we have

$$\ln n(T) - \ln n_0 = A(T - 0)$$

or

$$\ln \left[\frac{n(T)}{n_0}\right] = A T.$$

This equation is in the form $\ln x = AT$, where we have set $x = n(T)/n_0$. We can solve this for x by using the relation $e^{\ln x} = x$. Thus, $x = e^{\ln x} = e^{AT}$, and we have

$$\frac{n(T)}{n_0} = e^{AT}.$$

This expression describes the so-called exponential increase of population. Expressions of similar form describe many processes which are mathematically similar; for instance, the growth of money in banks due to interest, or the radioactive decay of atomic nuclei.

2. OSCILLATORY MOTION.

As a second example of a differential equation consider the motion of a particle in one dimension. It is sometimes possible to define the motion of the particle by a differential equation. For example, let x be the coordinate of the particle relative to the origin. Suppose we require that the position x of the particle satisfies the following differential equation,

$$\frac{d^2x}{dt^2} = -kx. \tag{1}$$

(This particular equation describes the motion of a pendulum, or of a particle suspended by a spring.)

The problem is to find out how x varies with time when it obeys this equation. This can be found by "solving" the differential equation. One of the most powerful means for solving differential equations is to guess a possible general form for the answer. Then this general form is substituted in the differential equation and one can both see that the equation is satisfied and determine any restrictions that should apply to the solution.

First, what is a promising guess as to a solution? Note that x must depend upon time in such a way that when it is differentiated twice with respect to time it reverses sign. But this is exactly what happens to the sine function since $\dfrac{d}{dx} \sin x = \cos x$,

and $\dfrac{d^2}{dx^2} \sin x = \dfrac{d}{dx} \cos x = -\sin x$ (frame 211). Therefore, let us try

$$x = A \sin (bt + c)$$

where A, b and c are undetermined constants.

This may be differentiated twice with respect to time with the result

$$\frac{dx}{dt} = Ab \cos (bt + c)$$

$$\frac{d^2x}{dt^2} = -Ab^2 \sin (bt + c).$$

If these relations are substituted in Equation (1), we have

$$-Ab^2 \sin (bt + c) = -kA \sin (bt + c)$$

The differentiated equation is then satisfied for all t provided

$$b^2 = k.$$

(Alternatively, the equation is satisfied by $A = 0$. However, this leads to a trivial result, $x = 0$, so we disregard this possibility.) Thus the solution is

$$x = A \sin (\sqrt{k}\, t + c)$$

Although the constant k is given by Equation (1), the constants A and c are arbitrary. If the position x and the velocity dx/dt were specified at some initial time, $t = 0$, the arbitrary constants could be determined.

Note that the solution we have found corresponds to x oscillating back and forth indefinitely between $x = A$ and $x = -A$. This type of oscillatory motion is characteristic of a pendulum or of a particle suspended by a spring, so that the original differential equation really appears to describe these systems.

Appendix B6

SUGGESTIONS FOR FURTHER READING

A good discussion of many of the functions which appear in this book is contained in

School Mathematics Study Group, Unit No. 21, *Elementary Functions*, Yale University Press, New Haven, Conn.

A clear discussion of limits can be found in the booklet by

G. B. Thomas, Jr., *Limits*, Addison-Wesley Publishing Co., Inc., Reading, Mass.

The following books are widely used as introductory texts in calculus:

M. H. Protter and C. B. Morrey, *Calculus With Analytic Geometry*, Addison-Wesley Publishing Co., Inc., Reading, Mass.

A. E. Taylor, *Calculus With Analytic Geometry*, Prentice-Hall, Inc., Englewood Cliffs, N. J.

G. B. Thomas, Jr., *Calculus and Analytic Geometry*, Addison-Wesley Publishing Co., Inc., Reading, Mass.

R. E. Johnson, E. L. Keokemeister, *Calculus With Analytic Geometry*, Allyn and Bacon, Inc., Boston.

A collection of trigonometric functions, logarithms and integrals is contained in the following:

The Handbook of Chemistry and Physics, Chemical Rubber Publishing Co., Cleveland, Ohio.

H. Dwight, *Tables of Integrals and Other Mathematical Data*, Macmillan Co., New York.

B. O. Peirce, *A Short Table of Integrals*, Ginn and Co., New York.

REVIEW PROBLEMS

This list of problems is for your benefit in case you want some additional practice. The problems are grouped according to chapter and section. Answers start on page 280.

CHAPTER I.

Section 3

Find the slope of the graphs of the following equations:

1. $y = 5x - 5$

2. $4y - 7 = 5x + 2$

3. $3y + 7x = 2y - 5$

Find the roots of the following:

4. $4x^2 - 2x - 3 = 0$

5. $(x^2 - 6x + 9) = 0$

Section 4

6. Show that $\sin \theta \cot \theta / \sqrt{1 - \sin^2 \theta} = 1$.

7. Show that $\cos \theta \sin (\frac{\pi}{2} + \theta) - \sin \theta \cos (\frac{\pi}{2} + \theta) = 1$.

8. What is: (a) $\sin 135°$, (b) $\cos \frac{7\pi}{4}$, (c) $\sin \frac{7\pi}{6}$?

9. Show that $\cos^2 \frac{\theta}{2} = \frac{1}{2} (1 + \cos \theta)$.

10. What is the cosine of the angle between any two sides of an equilateral triangle?

Section 5

11. What is $(-1)^{13}$?

12. Find $[(0.01)^3]^{-1/2}$.

13. Express $\log ([x^x]^x)$ in terms of $\log x$.

14. If $\log (\log x) = 0$, find x.

15. Is there any number for which $x = \log x$?

In the following 5 questions, make use of the log table below and the rules for manipulating logarithms.

x	$\log x$	x	$\log x$
1	0.00	5	0.70
2	0.30	7	0.85
3	0.48	10	1.00

Find

16. $\log \sqrt{10}$

17. $\log 21$

18. $\log \sqrt{14}$

19. $\log 300$

20. $\log 7^{3/2}$

CHAPTER II.

Find the following limits, if they exist:

21. $\lim\limits_{x \to 2} \dfrac{x^2 - 4x + 4}{x - 2}$

22. $\lim\limits_{\theta \to \pi/2} \sin \theta$

23. $\lim\limits_{x \to 0} \dfrac{x^2 + x + 1}{x}$

24. $\lim\limits_{x \to 1} \left[1 + \dfrac{(x + 1)^2}{(x - 1)}\right]$

25. $\lim\limits_{x \to 3} \left[(2 + x) \dfrac{(x - 3)^2}{x - 3} + 7\right]$

26. $\lim\limits_{x \to 1} \left[\dfrac{(x^2 - 1)}{x - 1}\right]$

27. $\lim\limits_{x \to \infty} \left(\dfrac{1}{x}\right)$

28. $\lim\limits_{x \to 0} \log x$

Section 3

29. What is the average velocity of a particle that goes forward 35 miles and backwards for 72 miles, during the course of one hour?

30. A particle always moves in one direction. Can its average velocity exceed its maximum velocity.

31. A particle moves so that its position is given by $S = S_0 \sin 2\pi t$ where S_0 is in meters, t is in hours. Find its average velocity from $t = 0$ to

    (a) $t = \dfrac{1}{4}$ hour,   (b) $t = \dfrac{1}{2}$ hour,

    (c) $t = 3/4$ hour,   (d) $t = 1$ hour.

32. Write an expression for the average velocity of a particle which leaves the origin at $t = 0$, whose position is given by $S = at^3 + bt$, where a and b are constants. The average is from $t = 0$ to the present.

33. Find the instantaneous velocity of a particle whose position is given by $S = bt^3$, where b is a constant, when $t - 2$.

Sections 5–8.

Find the derivative of each of the following functions with respect to its appropriate variable. a and b are constants.

34. $y = x + x^2 + x^3$

35. $y = (a + bx) + (a + bx)^2 + (a + bx)^3$

36. $y = (3x^2 + 7x)^{-3}$

37. $p = \sqrt{a^2 + q^2}$

38. $p = \dfrac{1}{\sqrt{a^2 + q^2}}$

39. $y = x^\pi$.

40. $f = \theta^2 \sin \theta$

41. $f = \dfrac{\sin \theta}{\theta}$

42. $f = (\sin \theta)^{-1}$

43. $f = (\sqrt{1 + \cos^2 \theta})^{-1}$

44. $f = \sin^2 \theta + \cos^2 \theta$

45. $y = \sin [\ln (x)]$

46. $y = x \ln x$

47. $y = (\ln x)^{-2}$

48. $y = x^x$
    (hint: what is $\ln y$? Use implicit differentiation, Appendix B3).

49. $y = a^{(x^2)}$

50. $f = \sin \sqrt{1 + \theta^2}$

51. $y = e^{-x^2}$

52. $y = \pi^x$

53. $y = \pi^{(x^2)}$

54. $f = \ln \sin \theta$

55. $f = \sin (\sin \theta)$

56. $f = \ln e^x$

57. $f = e^{\ln x}$

58. $y = \sqrt{1 - \sin^2 \theta}$

Section 9.

Evaluate each of the following:

59. Find $\dfrac{d^2}{d\theta^2} \cos a\theta$.

60. Find $\dfrac{d^n}{dx^n} e^{ax}$ (n is a positive integer).

61. $\dfrac{d^2}{dx^2} \sqrt{1 + x^2}$

62. $\dfrac{d^2}{d\theta^2} \tan \theta$

63. $\dfrac{d^3}{dx^3} x^2 e^x$

Section 10.

Find where the following functions have their maximum and/or minimum values. Either give the values of x explicitly, or find an equation for these values.

64. $y = e^{-x^2}$

65. $y = \dfrac{\sin x}{x}$

66. $y = e^{-x} \sin x$

67. $y = \dfrac{\ln x}{x}$

68. $y = e^{-x} \ln x$

69. Find whether y has a maximum or a minimum for the function given in question 64.

Section 11.

Find the differential df of each of the following functions.

70. $f = x$

71. $f = \sqrt{x}$

72. $f = \sin (x^2)$

73. $f = e^{\sin x}$ (hint: use chain rule)

CHAPTER III.

You may find Table 2 on page 285 helpful in doing the problems in this section.

Section 2.

Find the following indefinite integrals. (Omit the constants of integration.)

74. $\int \sin 2x \, dx$

75. $\int \dfrac{dx}{x + 1}$

76. $\int x^2 e^x dx$ (Try integration by parts.)

77. $\int xe{-}x^2 \, dx$

78. $\int \sin^2\theta \cos \theta \, d\theta$

Sections 3 and 4.

Evaluate the following definite integrals.

79. $\int_{-1}^{+1} (e^x + e^{-x}) \, dx$

80. $\int_{-\infty}^{\infty} \dfrac{dx}{a^2 + x^2}$

81. $\int_{-\infty}^{\infty} \dfrac{x dx}{\sqrt{a^2 + x^2}}$

82. $\int_{-\infty}^{0} x^2 e^x \, dx$ (Problem 76 may be helpful.)

83. $\int_{0}^{+\pi/2} \sin \theta \cos \theta \, d\theta$

84. $\int_{0}^{1} (x + a)^n \, dx$

85. $\int_{-1}^{+1} \dfrac{dx}{\sqrt{1 - x^2}}$

86. $\int_{-1}^{1} (x + x^2 + x^3) \, dx$

ANSWERS TO REVIEW PROBLEMS

1. 5

2. 5/4

3. −7

4. $(1 \pm \sqrt{13})/4$

5. 3,3 (roots are identical)

6. No answer

7. No answer

8. a) $\dfrac{\sqrt{2}}{2}$, b) $\dfrac{\sqrt{2}}{2}$, c) $-\dfrac{1}{2}$

9. No answer

10. $\dfrac{1}{2}$

11. −1

12. 1000

13. $x^2 \log x$

14. $x = 10$

15. No

16. 0.50

17. 1.33

18. 0.58

19. 2.48

20. 1.28

21. 0

22. 1

23. No limit

24. No limit

25. 7

26. 2

27. 0

28. No limit

29. − 37 mph.

30. No

31. a) $4\,S_0$ mph.,   b) 0 mph.,
    c) $-\dfrac{4}{3}\,S_0$ mph., d) 0 mph.

32. $at^2 + b$

33. $12b$

34. $1 + 2x + 3x^2$

35. $b + 2b\,(a + bx) + 3b\,(a + bx)^2$

36. $-3\,(3x^2 + 7x)^{-4}\,(6x + 7)$

37. $\dfrac{dp}{dq} = \dfrac{q}{\sqrt{a^2 + q^2}}$

38. $\dfrac{dp}{dq} = \dfrac{-q}{(a^2 + q^2)^{3/2}}$.

39. $\dfrac{dy}{dx} = \pi\,x^{(\pi - 1)}$

40. $\dfrac{df}{d\theta} = 2\theta \sin \theta + \theta^2 \cos \theta$

41. $\dfrac{df}{d\theta} = \dfrac{\cos \theta}{\theta} - \dfrac{\sin \theta}{\theta^2}$

42. $\dfrac{df}{d\theta} = -\dfrac{\cos \theta}{\sin^2 \theta}$

43. $\dfrac{df}{d\theta} = \dfrac{\cos \theta \sin \theta}{(1 + \cos^2 \theta)^{3/2}}$

44. $\dfrac{df}{d\theta} = 0$

45. $\dfrac{dy}{dx} = \dfrac{\cos [\ln (x)]}{x}$

46. $\dfrac{dy}{dx} = 1 + \ln x$

47. $\dfrac{dy}{dx} = \dfrac{-2}{x} (\ln x)^{-3}$

48. $\dfrac{dy}{dx} = x^x (1 + \ln x)$

49. $\dfrac{dy}{dx} = 2x \, a^{(x^2)} \ln a$

50. $\dfrac{\theta}{\sqrt{1 + \theta^2}} \cos \sqrt{1 + \theta^2}$

51. $-2xe^{-x^2}$

52. $\pi^x \ln \pi$

53. $2x\pi^{x^2} \ln \pi$

54. $\cot \theta$

55. $[\cos (\sin \theta)] \cos \theta$

56. 1

57. 1

58. $- \sin \theta$

59. $- a^2 \cos a\,\theta$

60. $a^n e^{ax}$

61. $\dfrac{1}{\sqrt{1 + x^2}} - \dfrac{x^2}{(1 + x^2)^{3/2}}$

62. $2 \sec^2 \theta \tan \theta$

63. $(6 + 6x + x^2) e^x$

64. $x = 0$

65. $x = \tan x \qquad (x = 0, \, \dots)$

66. $x = \arctan 1 = \dfrac{\pi}{4} \pm n\pi,$
$\qquad n = 0, 1, 2, \cdots$

67. $x = e \qquad (\ln x = 1)$

68. $\dfrac{1}{x} = \ln x$

69. Maximum

70. $df = dx$

71. $df = \dfrac{dx}{2\sqrt{x}}$

72. $df = 2x \cos (x^2) \, dx$

73. $df = \cos x e^{\sin x} \, dx$

74. $\dfrac{-1}{2} \cos 2x$

75. $\ln (x + 1)$

76. $x^2 e^x - 2xe^x + 2e^x$

77. $-\dfrac{1}{2} e^{-x^2}$

78. $\dfrac{1}{3} \sin^3 \theta$

79. $2 \left(e - \dfrac{1}{e}\right)$

80. $\dfrac{\pi}{a}$

81. 0

82. 2

83. $\dfrac{1}{2}$

84. $\dfrac{(1 + a)^{n+1} - a^{n+1}}{(n + 1)}$

85. π

86. $\dfrac{2}{3}$

TABLES

Table 1

DERIVATIVES

The differentiation formulas are listed below. References to the appropriate frames are given. In the following expressions ln x is the natural logarithm or the logarithm to the base e; u and v are variables that depend on x; w depends on u which in turn depends on x; and a and n are constants. All angles are measured in radians.

Frame

1. $\dfrac{da}{dx} = 0$ 172

2. $\dfrac{d}{dx}\, ax = a$ 174

3. $\dfrac{dx^n}{dx} = nx^{n-1}$ 180

4. $\dfrac{d}{dx}\,(u + v) = \dfrac{du}{dx} + \dfrac{dv}{dx}$ 186

5. $\dfrac{d}{dx}\,(uv) = u\,\dfrac{dv}{dx} + v\,\dfrac{du}{dx}$ 189

6. $\dfrac{d}{dx}\left(\dfrac{u}{v}\right) = \dfrac{1}{v^2}\left[v\,\dfrac{du}{dx} - u\,\dfrac{dv}{dx}\right]$ 202

7. $\dfrac{dw}{dx} = \dfrac{dw}{du}\,\dfrac{du}{dx}$ 194

8. $\dfrac{du^n}{dx} = n\,u^{n-1}\,\dfrac{du}{dx}$ From Eqs. (2) and (7).

9. $\dfrac{d\ln(x)}{dx} = \dfrac{1}{x}$ 230

10. $\dfrac{d\log_{10} x}{dx} = \dfrac{1}{x}\,\log_{10} e$ 234

11. $\dfrac{de^x}{dx} = e^x$ 239

12. $\dfrac{da^x}{dx} = a^x \ln(a)$ 238

13. $\dfrac{du^v}{dx} = vu^{v-1}\dfrac{du}{dx} + u^v \ln(u)\dfrac{dv}{dx}$

14. $\dfrac{d\sin x}{dx} = \cos x$ 210

15. $\dfrac{d\cos x}{dx} = -\sin x$ 211

16. $\dfrac{d\tan x}{dx} = \sec^2 x$ 212

17. $\dfrac{d\sec x}{dx} = \sec x \tan x$ 213

18. $\dfrac{d\cot x}{dx} = -\csc^2 x$

19. $\dfrac{d\arcsin x}{dx} = \dfrac{1}{\sqrt{1-x^2}}$ (Appendix B4)

20. $\dfrac{d\arccos x}{dx} = \dfrac{-1}{\sqrt{1-x^2}}$ (Appendix B4)

21. $\dfrac{d\arctan x}{dx} = \dfrac{1}{1+x^2}$ (Appendix B4)

22. $\dfrac{d\operatorname{arccot} x}{dx} = \dfrac{-1}{1+x^2}$ (Appendix B4)

Table 2

INTEGRALS

The following list of integrals from frame 307 is reproduced here for convenience. In the list u and v are variables that depend on x; w is a variable that depends on u which in turn depends on x; a and n are constants; and the arbitrary integration constants are omitted for simplicity.

1. $\int a\, dx = ax$

2. $\int a\, f(x)\, dx = a \int f(x)\, dx$

3. $\int (u + v)\, dx = \int u\, dx + \int v\, dx$

4. $\int x^n\, dx = \dfrac{x^{n+1}}{n + 1} \quad n \neq -1$

5. $\int \dfrac{dx}{x} = \ln x$

6. $\int e^x\, dx = e^x$

7. $\int e^{ax}\, dx = e^{ax}/a$

8. $\int b^{ax}\, dx = \dfrac{b^{ax}}{a \ln b}$

9. $\int \ln x\, dx = x \ln x - x$

10. $\int \sin x\, dx = -\cos x$

11. $\int \cos x\, dx = \sin x$

12. $\int \tan x\, dx = -\ln \cos x$

13. $\int \cot x\, dx = \ln \sin x$

14. $\int \sec x\, dx = \ln (\sec x + \tan x)$

15. $\int \sin x \cos x\, dx = \dfrac{1}{2} \sin^2 x$

16. $\int \dfrac{dx}{a^2 + x^2} = \dfrac{1}{a} \arctan \dfrac{x}{a}$

17. $\displaystyle\int \frac{dx}{\sqrt{a^2 - x^2}} = \arcsin \frac{x}{a}$

18. $\displaystyle\int \frac{dx}{\sqrt{x^2 \pm a^2}} = \ln \left[x - \sqrt{x^2 \pm a^2} \right]$

19. $\displaystyle\int w\,(u)\,dx = \int w\,(u)\,\frac{dx}{du}\,du$

20. $\displaystyle\int u\,dv = uv - \int v\,du$

INDEX

References are by page number. A separate index of symbols can be found on page 292.

INDEX OF SYMBOLS

References are by page number